→IN

PHILOSOPHY
OF SCIENCE

ZIAUDDIN SARDAR & BORIN VAN LOON

Published in the UK in 2011 by Icon Books Ltd, Omnibus Business Centre, 39–41 North Road, London N7 9DP
email: info@iconbooks.co.uk
www.introducingbooks.com

Sold in the UK, Europe, South Africa and Asia by Faber & Faber Ltd, Bloomsbury House, 74–77 Great Russell Street, London WC1B 3DA or their agents

Distributed in the UK, Europe, South Africa and Asia by TBS Ltd, TBS Distribution Centre, Colchester Road, Frating Green, Colchester CO7 7DW

This edition published in Australia in 2011 by Allen & Unwin Pty Ltd, PO Box 8500, 83 Alexander Street, Crows Nest, NSW 2065

Previously published in the UK and Australia in 2002

This edition published in the USA in 2011 by Totem Books
Inquiries to: Icon Books Ltd, Omnibus Business Centre, 39–41 North Road, London N7 9DP, UK

Distributed to the trade in the USA by Consortium Book Sales and Distribution
The Keg House,
34 Thirteenth Avenue NE, Suite 101, Minneapolis, MN 55413-1007

Distributed in Canada by Penguin Books Canada, 90 Eglinton Avenue East, Suite 700, Toronto, Ontario M4P 2Y3

ISBN: 978-184831-296-8

Text copyright © 2002 Ziauddin Sardar
Illustrations copyright © 2002 Borin Van Loon

The author and artist have asserted their moral rights.

Originating editor: Richard Appignanesi

No part of this book may be reproduced in any form, or by any means, without prior permission in writing from the publisher.

Printed by Gutenberg Press, Malta

The Nature of the Beast

Our world is shaped and driven by science. Almost every benefit of modern life – from antibiotics to computers, our understanding of human evolution to our ability to land a satellite on Saturn – is a product of science. For most people, progress is simply another term for advances in scientific knowledge and benefits derived from new discoveries of science.

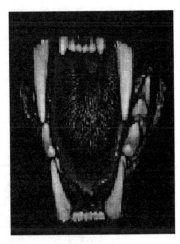

But what exactly is this perpetual engine of progress?

While the benefits of science are easy to see, science itself is anything but easy to define.

Is Science "Absolute Objectivity"?

Until quite recently, Western tradition saw science as the quest for objective knowledge of nature and reality. Scientists were regarded as quasi-religious supermen, heroically battling against all odds to discover the truth.

And the truths they wrestled out of nature were said to be <u>absolute</u> ...

... objective, value-free and universal.

As one sociologist in the 1940s described it, science reflects the character of nature itself: "The stars have no sentiments, the atoms no anxieties which have to be taken into account. Observation is objective with little effort on the part of the scientist to make it so."

Or, as J.D. Bernal (1901–71), the radical historian of science, put it ...

Science is all about rationality, universalism and disinterestedness.

Do We Trust Scientists?

But this picture of truth-loving and truth-seeking scientists working for the benefit of humanity is rather at odds with the public conception of science and scientists. Most people are not "anti-science". We recognize the potential that science has for making our lives healthier and easier.

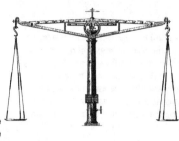

But recent research has shown that most people do not trust scientists and are concerned with potential harmful side-effects of science.

Scientists are seen by the public not as disinterested "truth-seekers" but as narrow-minded compulsives concerned with their own fame and fortune.

Office of Science & Technology and The Wellcome Trust

Science And The Public
A Review Of Science Communication And Public Attitudes To Science In Britain

London 2000

The view of the scientists we find in popular literature and film is even more scathing.

Dr Henry Frankenstein of Mary Shelley's *Frankenstein* (1818) is not the monster, but ...

> ... a man of science who sought to create a man after his own image – without reckoning upon God.

In Robert Louis Stevenson's *Dr Jekyll and Mr Hyde* (1886), Jekyll is a restless young scientist who discovers a concoction that turns him into his alter ego ...

> ... the repellent and murderous Mr Hyde.

In H.G. Wells' *The Island of Doctor Moreau* (1896), a scientist develops mutant life-forms that live in pain and misery ...

> ... we violently revolt against our creator.

> I used to be a contender...

In the classic film *Dr Strangelove* (1964), the title character, played by Peter Sellers, is a paraplegic Nazi scientist...

...who is miraculously cured once the world has been plunged into a nuclear Armageddon.

The Boys from Brazil (1978) shows scientists as evil Nazis hell-bent on recreating a race of Hitlers.

In *Batman and Robin* (1997), both villains are scientists:

...the evil Mr Freeze...

...and the misguided Miss Poison Ivy.

Why do the popular perceptions of science and scientists differ so radically from the scientists' own self-image as brilliant pioneers deserving of admiration, funding and blind trust? Perhaps because, apart from bringing benefits, science has also posed serious threats to humanity.

Science has given us the bomb, as well as biological and chemical weapons of mass destruction.

It introduced the spectre of eugenics and has brought us to the brink of human cloning.

The by-products of science, such as nuclear waste and chemical pollution, are destroying ecosystems on local, regional and global scales. So, science brings us benefits as well as costs. Perhaps it was in an effort to present a more deflated image of science that the Nobel Prize physicist **Lord Rutherford** (1871–1937) said:

Science is what scientists do.

What Do Scientists Actually Do?

Here are some examples of the negative things that scientists actually do, as reported by the media.

The Independent newspaper, Section 2, 26 January 1995, "They Shoot Pigs Don't They?" reported:

In Porton Down research establishment in England, scientists have been using live animals to test body armour. The animals were strapped on to trolleys and subjected to blasts at either 600 or 750mm from the mouth of the explosively driven shock tube. Initially, monkeys were used in these experiments, but scientists later switched to shooting pigs. The animals were shot just above the eye to investigate the effects of high-velocity missiles on brain tissue.

Time magazine, January 1994; also Chip Brown, "The Science Club Serves its Country", *Esquire*, December 1994 reported:

In the United States in the late 1940s, teenage boys were fed radioactive breakfast cereal, middle-aged mothers were injected with radioactive plutonium and prisoners had their testicles irradiated – all in the name of science, progress and national security. These experiments were conducted through to the 1970s.

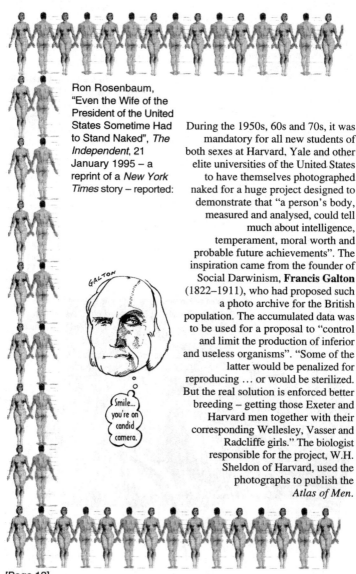

Ron Rosenbaum, "Even the Wife of the President of the United States Sometime Had to Stand Naked", *The Independent*, 21 January 1995 – a reprint of a *New York Times* story – reported:

GALTON

Smile... you're on candid camera.

During the 1950s, 60s and 70s, it was mandatory for all new students of both sexes at Harvard, Yale and other elite universities of the United States to have themselves photographed naked for a huge project designed to demonstrate that "a person's body, measured and analysed, could tell much about intelligence, temperament, moral worth and probable future achievements". The inspiration came from the founder of Social Darwinism, **Francis Galton** (1822–1911), who had proposed such a photo archive for the British population. The accumulated data was to be used for a proposal to "control and limit the production of inferior and useless organisms". "Some of the latter would be penalized for reproducing ... or would be sterilized. But the real solution is enforced better breeding – getting those Exeter and Harvard men together with their corresponding Wellesley, Vasser and Radcliffe girls." The biologist responsible for the project, W.H. Sheldon of Harvard, used the photographs to publish the *Atlas of Men*.

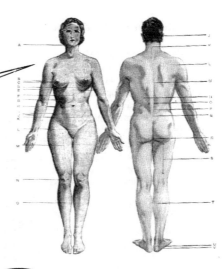

"These revelations cast science in a radically different perspective."

What scientists actually do has been extensively dissected by historians of science, examined by sociologists and anthropologists of science, analysed by philosophers of science, and scrutinized by feminist and non-Western scholars.

"This work has produced a different set of definitions and explanations for science ..."

"One that challenges the scientists' own view of science as an objective adventure that stands above all concerns of culture and values."

Definitions of Science

Most critics now see science as an organized, institutionalized and industrialized venture. It requires huge funding, large, sophisticated and expensive equipment and hundreds of scientists working on minute problems.

The prospects for technological application – usually for profit – determine the choices of which scientific projects and fields will be funded ...

... and which will be starved.

As knowledge and power have coalesced, knowledge itself has been corrupted and become an instrument of social control and corporate domination.

Here are some other definitions of science.

Steve Fuller, Professor of Sociology, Warwick University

> Science is a sexist and chauvinist enterprise that promotes the values of white, middle-class males.

> Science is the systematic pursuit of knowledge, regardless of subject matter. What is sociologically most interesting about science is that it sets the standard by which the rest of society is legitimated. This standard often goes by the name of "rationality", "objectivity", or simply "truth". When we use these words, we imply that the standard of legitimation is, at least in principle, available to everyone in society. This is simply not the case. The opposite of science is not ideology or technology, but expertise and intellectual property which imply that knowledge is privatised to a select group of knowledge-producers and owners.

Sandra Harding, feminist scholar of science

The Golem of Science

> Science is a golem. A golem is a creature of Jewish mythology. It is a humanoid made by man from clay and water, with incantations and spells. It is powerful. It grows a little more powerful every day. It will follow orders, do your work, and protect you from the ever threatening enemy. But it is clumsy and dangerous. Without control, a golem may destroy its masters with its failing vigour ... since we are using a golem as a metaphor for science, it is also worth noting that in the mediaeval tradition the creature of clay was animated by having the Hebrew word "EMETH", meaning truth, inscribed on its forehead – it is truth that drives it on. But this does not mean it understands the truth – far from it.

Harry Collins and Trevor Pinch,
sociologists of science

The Contested Territory of Science

"Science is a theology of violence. It performs violence against the subject of knowledge, against the object of knowledge, against the beneficiary of knowledge and against knowledge itself."

Ashis Nandy, Indian cultural theorist

"Science is the new entrenched state religion in America."

Vine Deloria Jr., Lakota Indian activist and Professor of American Indian Studies, University of Colorado

All of these different definitions and perceptions of science tell us one thing for certain: Science is a contested territory.

The various claims and counter-claims about the nature of science – all containing some aspect of truth – reveal science to be a highly complex and multi-layered activity. No single and simple description of science can reveal its basic nature. No romantic ideal can describe its real character. No sweeping generalization can uncover its real dimensions.

Do Scientists Understand Science?

Until now, scientists have had little or no understanding of how science actually works in practice. Scientists have misunderstood science in a number of important ways.

- They have had a rather romantic notion of scientific method which, they are taught to believe, magically produces neutral, value-free and universal Truth statements.

- They have thought that they are operating in an autonomous environment protected by state funding. In reality, funding for science increasingly comes from corporations and foundations with vested interests in certain research agendas.

- They have thought that the sole purpose of research is to advance human understanding and knowledge. In reality, science is driven by military interests, the need for corporations to make profit, and those concerns of the public that cannot be politically ignored.

- They have tended to believe that science can be pursued for its own sake. It should remain esoteric in content, accountable only to itself, with no concern for social or cultural issues, *and* be publicly funded. But that's not how democracies work.

- They have tended to presume – wrongly – that if the public were to have more technical knowledge of science, it would accept what they say implicitly. The public is often concerned with questions of ethics and policies and risks and safety – topics about which scientists know very little.

> Given that scientists have knowledge only about their specialist field of activity, it is not surprising that experts from other disciplines – philosophy, history, sociology – have tried to fill a void in our knowledge and action left open by scientists.

> This is where Science Studies comes in ...

Science Studies in the 1960s

Science Studies itself began in the late 1960s largely at the instigation of historians and philosophers of science, radical scholars, environmentalists and concerned scientists who had become disillusioned with science's incorporation into the military-industrial complex. Degree programmes were started to integrate "science, technology and society".

> These were usually run from science and engineering faculties in liberal arts settings. Their orientation tended to be critical of the status quo.

Kent State University 1970

> Science Studies fostered such counter-cultural trends as "small is beautiful", radical science, and movements concerned with the empowerment of women and ethnic minorities.

> You may be small, but you're perfectly formed.

Diverse Critical Approaches

The loose amalgam of critical approaches to science went under a number of different rubrics, including ...

- Science, Technology and Societies Studies
- Science Policy Studies
- Social Studies of Science
- Science, Technology and Development Studies
- Science, Technology and Culture Studies
- Sociology of Science and Technology

So... ... radical science... flower power... revolutionary student politics... integrating science with society... liberation of minorities... Marxism... Marcusianism... ... What was that about empowerment of women?

Outside the academy, Science Studies was championed by the environmental movement, "Science for the People" groups, and various Marxist and Socialist critics of science.

Who's he?

Hang on. Here comes a bloke with a placard.

And who's he?

Jerome Ravetz (b. 1929) → Philosopher of Science (and many other things)

In all cases, the critical mission of Science Studies was to reform science <u>in society.</u>

I think it says on the next page!

A Growth Industry

In Britain, the first self-declared school of Science Studies was the Sociology of Scientific Knowledge (or the "Strong Programme") established in the 1960s at Edinburgh University. It was a product of Labour Prime Minister **Harold Wilson** (1916–95) …

Our efforts must be to bridge the "two cultures" – the sciences and humanities – in a period when science is becoming increasingly important at all levels of policy-making.

Portrait of the author as a young man

By the 1970s, university expansion had enabled Science Studies to become a discipline in its own right.

THE WHITE HEAT OF TECHNOLOGY

Science Studies started to adopt the trappings of the sciences it studied, including specialist journals, professional societies, and claims to disciplinary autonomy, based on the accumulation of "case studies".

Criticism from the "Low Church"

During the 1980s, works like Ashis Nandy's *Science, Hegemony and Violence* (1988), and …

… my own <u>The Revenge of Athena: Science, Exploitation and the Third World</u> (1988), exposed the racial and political economy of science.

By the end of the Cold War, Science Studies had established itself as a respectable discipline.

While it became a major irritant for the scientific community, some scientists themselves began to see Science Studies as a vehicle for defending and improving their own practices.

Comparing the Radical Origins

One way to appreciate the transition of Science Studies from radical scholarly subject to professionalized discipline is to compare the contents of two important Science Studies handbooks. When it was first published in 1977, the book:

I contributed a chapter on "Science policy and developing countries".

I contributed a chapter on "Criticisms of Science".

SCIENCE, TECHNOLOGY AND SOCIETY
A Cross-Disciplinary Perspective

Edited by
Ina Spiegel-Rösing and Derek de Solla Price

under the aegis of the
International Council for Science Policy Studies

... was seen as a ground-breaking reader.

CONTENTS

Preface   1

PART ONE THE NORMATIVE AND PROFESSIONAL CONTEXTS

1 The Study of Science, Technology and Society (SSTS): Recent Trends and Future Challenges
I. Spiegel-Rösing   7

Institutional Development and 'Regionalizations'
Origins of the Field · Institutional Development · Sociopolitical Regionalization · Cognitive Regionalization and the Place of 'Scientometrics'

Tendencies and Deficiencies
Some Recent Research Tendencies in the Field · Humanistic Tendencies · Relativistic Tendencies · Reflexive Tendencies · De-simplifying Tendencies · Normative Tendencies · Deficiencies · Rhetoric Pathos · Fragmentation · Lack of Comparative Research · The Bigness and Hardness Bias · Future Research

SSTS: Its Audiences, Critics, Tasks
Knowledge Production and Its Impurities · Application and Its Impracticabilities · Ethics of Science and Technology in Society

Bibliography

2 Science Policy Studies and the Development of Science Policy
Jean-Jacques Salomon   43

Introduction
The Prehistory of Science Policy
The Rise of Science Policy
The Age of Pragmatism
The Age of Questioning

3 Criticisms of Science
J. R. Ravetz

An Example from Classical Civilization
Criticism in the Scientific Revolution
The Romantic Challenge and Its Descendants
Modern Radical Criticisms of Science
Science Policy Studies: From Publicity to Politics
Doubts Among the Scholars
Personal Interpretation
Bibliography

4 *M. J. Mulkay*   93

Introduction: Pure and Applied Research
The Norms of Pure Research
The Distribution of Professional Rewards
Social Exchange and Social Control
The Formal and Informal Organization of Pure Research
Problem Networks and Processes of Communication
The Development of Research Networks
Competition and Secrecy
Intellectual Resistance
The Future of the Pure Research Community
Social Characteristics of Applied Research
Summary
Evaluation
Bibliography

5 Changing Perspectives in the Social History of Science
Roy MacLeod   149

Introduction
The Social History of Science and the Internalist/Externalist Debate
The Changing Scope of the Social History of Science
Epistemological Obstacles and Externalist Goals
Bibliography

6 Conditions of Technological Development
E. Layton   197

Introduction: Science Policy and Technology
Definitions of Terms · Technology and Technique

Measurement of the Growth of Technology and Technique
Production Functions · Patent Statistics · Indirect Social Costs and Benefits of Technological Change · The Study of Innovations · The Flow of Information · Embodiment of Science in Technology

The Sources of Technical Development:
Linear-Sequential Models
Necessary and Sufficient Conditions of Technological Change · Conflicts Among Innovation Studies

The Role of Science in Historical Perspective
The History of Technology · Interactions of Science and Technology · The Technological 'Sciences'

The Institutionalization of Applied Science
The Interdisciplinary Research Laboratory · Convergence in Research · Government R & D · Industrial R & D · Technological Sophistication and Social Need

Changing Conceptions of Technology
Bibliography

7 Economics of Research and Development
C. Freeman
223

Introduction
Definitions and Conceptual Framework
  Technology · Technical Change
  · Inventions, Innovations and Diffusion
  · Research and Experimental Development
  · Summary of Definitions
A Historical Review of Economic Thought on Technical Change
  Adam Smith and the Classical Economists · Malthus · Marx
  · Neoclassical Economics · Schumpeter
  · The Economics of Oligopoly and Galbraith
Some Recent Empirical Research
  R & D Statistics and Technological Change · The Sources of Invention and Innovation · Research, Innovation and Size of Firm · Uncertainty, Management of Innovation, and Theory of the Firm
  · Project Evaluation, Cost Benefit, Programming and Technology Assessment
  · Conclusions and Future Research
Bibliography

8 Psychology of Science
R. Fisch
277

Introduction
General Studies on Scientists and Technologists
Motives, Norms and Values, Political Attitudes
  Motives · Norms and Values · Political Attitudes
Psychological Aspects in the Development of Scientists
  Socialization Processes · Scientific Career · Mobility · Creativity and Productivity · Creativity, Productivity and its Criteria · Creative Scientists
  · Sex Differences · Environmental Conditions
Conclusion
Bibliography

9 Models for the Development of Science
Gernot Böhme
319

Introduction
Models for the Development of Science
Phases of the Development of Science: Kuhn's Theory of Normal and Revolutionary Science
Continuity in the Development of Science
Evolutionary Development Models
  Historical Change of Developmental Models
The Interaction Between Scientific Development and Technical Development
Marxist Concepts of the Development of Science
The State of the Art: Future Prospects
Bibliography

PART THREE SCIENCE POLICY STUDIES: THE POLICY PERSPECTIVE

10 Scientists, Technologists and Political Power
Sanford A. Lakoff
355

Historical Evolution: Past and Present
The Tradition that Science is Politically Neutral
The End of Neutrality and Internationalism
World War II and the Development of the Atomic Bomb
The Atomic Arms Race and the 'Scientists' Movement'
The Expanding Social Role of the Scientist and Technologist
The Specter of Technocracy
Political Dimensions: The Political Characteristics of Scientists and Technologists
The Political Functions of Scientists and Technologists
  As Advocates of Support · As Advisers · As Adversaries
Scientists, Technologists and Social Responsibility
Epilogue: Knowledge and Power in Scholarly Perspective
Bibliography

11 Technology and Public Policy
D. Nelkin
393

Introduction
The Use of Science and Technology
  Allocation of Resources · Strategies
  · Problems of Utilization

Impact of Science and Technology
  Areas of Impact · Policy Importance
Control
  Participatory Controls · Reactive Controls
  Anticipatory Controls
Conclusion
Bibliography

12 Science, Technology and Military Policy
Harvey M. Sapolsky
443

Introduction
Science, Technology and War
Military R & D as a National Priority
The Nature of Modern Weapons
The Organization of Military R & D Efforts
The Effect on the Military: Science and Warfare
Arms Control
Bibliography

13 Science, Technology and Foreign Policy
Brigitte Schroeder-Gudehus
473

Introduction: Historical Perspective
Science and Technology as Power Factors
  Goals and Instruments of Foreign Policy
  · International Cooperation and Transnational Actors
The Process of Foreign Policy Making: Adjustments, Gaps and Barriers
Political Dimensions of the International Scientific Community
Bibliography

14 Science, Technology and the International System
Eugene B. Skolnikoff
507

General Effects of Scientific and Technological Development: Introduction
Five General International Effects of Technology
  Interdependence · The Meaning of Warfare · New Patterns of Interactions and New Actors · Rich and Poor
  · Domestic Policy Processes: Foreign Policy and Feedback to Science Policy

Seven Specific Issue-Areas
  Food and Population · Energy · Atomic Energy · Environment · Oceans · Outer Space · Technology and Trade, Multinationals, Transfer of Technology
Approaches
Conc...

15 Science Policy and Developing Countries
Ziauddin Sardar and Dawud G. Rosser-Owen

Scope and Terminology
  Introduction: A Three Faction World?
  The Concept of the Occident · The Development Continuum · Conspicuous Technology

Historical Perspective
  Planning for Development: The Conventional Views · Internal Sources of Income · Foreign Aid · Foreign Loans · Pearson's Report

Some Aspects of Development
  The Multidimensional Process · Social Capital · The Traditional Background · Educational Systems · Education and Training · Agriculture and Land Reform · Industrialization and Manpower Problems

Some Recent Trends
  New Theories of Underdevelopment · Technology Transfer · The Chinese Model of Development · Alternative Technology · The UNCTAD Meeting

Contents

Foreword ix
Introduction xi

Part I. Overview

1. Reinventing the Wheel 3
   David Edge

Part II. Theory and Methods 25

2. Four Models for the Dynamics of Science 29
   Michel Callon
3. Coming of Age in STS: Some Methodological Musings 64
   Gary Bowden
4. The Origin, History, and Politics of the Subject Called "Gender and Science": A First Person Account 80
   Evelyn Fox Keller
5. The Theory Landscape in Science Studies: Sociological Traditions 95
   Sal Restivo

Part III. Scientific and Technical Cultures 111

6. Science and Other Indigenous Knowledge Systems 115
   Helen Watson-Verran and David Turnbull
7. Laboratory Studies: The Cultural Approach to the Study of Science 140
   Karin Knorr Cetina
8. Engineering Studies 167
   Gary Lee Downey and Juan C. Lucena
9. Feminist Theories of Technology 189
   Judy Wajcman
10. Women and Scientific Careers 205
    Mary Frank Fox

Part IV. Constructing Technology 225

11. Sociohistorical Technology Studies 229
    Wiebe E. Bijker
12. From "Impact" to Social Process: Computers in Society and Culture 257
    Paul N. Edwards
13. Science Studies and Machine Intelligence 286
    H. M. Collins
14. The Human Genome Project 302
    Stephen Hilgartner

Part V. Communicating Science and Technology 317

15. Discourse, Rhetoric, Reflexivity: Seven Days in the Library 321
    Malcolm Ashmore, Greg Myers, and Jonathan Potter
16. Science and the Media 343
    Bruce V. Lewenstein
17. Public Understanding of Science 361
    Brian Wynne

Part VI. Science, Technology, and Controversy 389

18. Boundaries of Science 393
    Thomas F. Gieryn
19. Science Controversies: The Dynamics of Public Disputes in the United States 444
    Dorothy Nelkin
20. The Environmental Challenge to Science Studies 457
    Steven Yearley
21. Science as Intellectual Property 480
    Henry Etzkowitz and Andrew Webster
22. Scientific Knowledge, Controversy, and Public Decision Making 506
    Brian Martin and Evelleen Richards

Part VII. Science, Technology, and the State 527

23. Science, Government, and the Politics of Knowledge 533
    Susan E. Cozzens and Edward J. Woodhouse
24. Politics by the Same Means: Government and Science in the United States 554
    Bruce Bimber and David H. Guston
25. Changing Policy Agendas in Science and Technology 572
    Aant Elzinga and Andrew Jamison
26. Science, Technology, and the Military: Relations in Transition 598
    Wim A. Smit
27. Science and Technology in Less Developed Countries 627
    Wesley Shrum and Yehouda Shenhav
28. Globalizing the World: Science and Technology in International Relations 652
    Vittorio Ancarani

References 671
Index 774
About the Authors 809

Why is Science Studies Important?

Science Studies is definitely *not* important as simply another empirical academic discipline or branch of sociology. Its importance lies solely as a vehicle for surveying, criticizing and transforming our knowledge practices more generally.

Science Studies' most important lesson is that science has been generally blind to the <u>social character</u> of its own practices.

This has been the major source of its problems in relating to society.

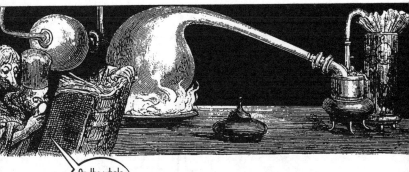

- introduce a discourse of values in the practice and operation of science
- open the practice of science to democratic accountability, especially its decision-making processes and power structures
- interrogate the kinds of questions science asks, what type of solutions it seeks, and the implicit assumptions that order its operations and practices
- examine the in-built gender and racial biases in the process of science
- seek out the consequences of the mono-cultural matrix that powers science, exposing the possibilities of multiple, as well as multi-cultural, means of doing science

The Great March

From now on, we move from one Great Scientist to another in a great chain of scientific being – and science vanquishes ignorance, superstition and dogma.

Philosopher and mathematician **René Descartes** (1596–1650) showed that the rainbow is not a heavenly signal of peace.

Rather, it can be explained in terms of what happens to rays of light when they encounter raindrops. I, Descartes, also showed why rainbows have the shape of a circle around the sun, and are always at the same distance from it.

Planet of the Apes

Charles Darwin (1809–82) gave us the bad news that our Adam-and-Eve origin was just a fable.

Science in the Killing-fields

But after the First World War, this conventional history of science became slightly problematic. A great scientist in Germany, **Fritz Haber** (1868–1934), Nobel Prize winner for Chemistry, invented poison gas.

Used against other Europeans, and not just "natives", this seemed a horrible perversion of science.

Then the war against Japan ended with the atomic bomb. However necessary it may have been to shorten the war, it seemed to raise powers of a supernatural kind. With the development of the H-bomb and intercontinental ballistic missiles, the fruits of science were able to destroy us all.

The anti-nuclear movement with its "peace lollipop" was a constant reminder that science could go horribly wrong.

Can Scientists Make Mistakes?

A Question of Paradigms

On one very hot day, Kuhn realized that Aristotle had not been getting the "wrong answer" to Galileo's problem.

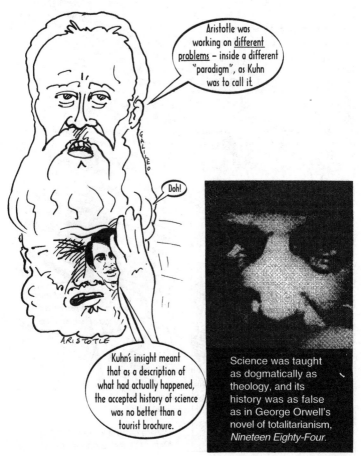

Aristotle was working on <u>different problems</u> – inside a different "paradigm", as Kuhn was to call it.

Doh!

Kuhn's insight meant that as a description of what had actually happened, the accepted history of science was no better than a tourist brochure.

Science was taught as dogmatically as theology, and its history was as false as in George Orwell's novel of totalitarianism, *Nineteen Eighty-Four*.

The triumphalist image of science had fallen off the wall, and, like Humpty-Dumpty, it could never be put back together again. Both history and philosophy of science, as well as Science Studies, played their part in bringing science down to human dimensions.

Where do we begin our story of science's fall from grace?

Well, we have to begin somewhere. So let's begin with the Vienna Circle.

The Vienna Circle: Logical Positivism

Established in the 1920s, the Vienna Circle was an influential school of philosophy of science. At its height, it had about three dozen members, drawn from natural and social sciences, logic and mathematics. Its leading members, **Rudolf Carnap** (1891–1970) and **Otto Neurath** (1882–1945), saw it as a means of advancing anti-clerical and Socialist ideas. The Circle's first publication was its manifesto: *The Scientific Conception of the World* (1929).

The position of the Circle, upheld in its journal <u>Erkenntnis</u> – <u>Knowledge</u>, later called <u>The Journal of Unified Sciences</u> – asserts that metaphysics and theology are meaningless ...

They consist of proposition that cannot be verified.

Its own doctrine, known as **logical positivism**, conceived philosophy as purely analytical, based on formal logic, and the only legitimate component of scientific discourse.

The Circle's Influence

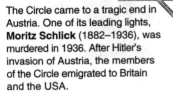

The Circle came to a tragic end in Austria. One of its leading lights, **Moritz Schlick** (1882–1936), was murdered in 1936. After Hitler's invasion of Austria, the members of the Circle emigrated to Britain and the USA.

> In the 1940s, the Circle's ideas became widely known, contributing to the emergence of the modern analytic philosophy of science.

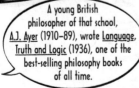

> A young British philosopher of that school, <u>A.J. Ayer</u> (1910–89), wrote <u>Language, Truth and Logic</u> (1936), one of the best-selling philosophy books of all time.

But the *political* origins of positivism were forgotten – or suppressed. It just seemed to be a dry doctrine proclaiming the infallibility of science.

Karl Popper's "falsifiability" Theory

Karl Popper (1902–94) was loosely associated with the Vienna Circle. He became one of the most innovative post-war philosophers of science. His theory of "falsifiability" undermined the then dominant view that accumulated experience leads to scientific hypothesis – dubbed "verification" by the Vienna Circle.

> Popper suggested that freely conjectured hypotheses precede and are tested against experience.

> "Falsifiability" – the fact that a scientific theory can be proved false by a single contrary incident – is the genuine demarcation between science and non-science.

Against Induction

Popper developed his ideas on the nature of scientific procedure in *The Logic of Scientific Discovery* (German original, 1934; translation, 1959). He disagreed with traditional beliefs about "induction" – general conclusions drawn from a set of given premises – which is the basis for all generalization in science.

> The models of "the language of science" which philosophers construct have nothing to do with the language of modern science.

> No number of particular instances, for example of A being a B, can establish a universal principle that all A's are B's. The "universal principle" can be refuted by a single instance of an A not being a B.

> This is the only certain inference in science. Popper used it to demarcate science from non-science or pseudo-science.

For Popper, there is no such thing as the final truth in science. Instead, scientific progress is achieved by *Conjectures and Refutations* (the title of his book of essays, published in 1963). This self-critical spirit was the essence of science for Popper.

Thomas Kuhn's Revolution

Thomas Samuel Kuhn (1922–96) is one of the most important scholars in Science Studies. Born in Cincinnati, Ohio, he studied physics at Harvard University and went on to do graduate studies in theoretical physics.

But before completing my dissertation, I decided to change to history of science ...

In 1962, Kuhn published *The Structure of Scientific Revolutions*, which has now become a decisive work on the nature of science in the 20th century. It is the source of such buzzwords as "paradigm", "revolutionary science" and (indirectly) "post-normal science".

The Structure of Scientific Revolutions

Kuhn explores big themes in science. He wants to know what science is really like – in its actual practice – in a concrete and empirical way. He suggests that far from discovering truth, scientists actually solve puzzles within established world-views.

This must be what they call a Rubric's Cube.

I used the term "paradigm" to describe the belief system that underpins puzzle-solving in science.

The term "paradigm" suggests that some accepted examples of actual scientific practice – which have produced theory, law, application and instrumentation – provide models from which spring particular coherent traditions of scientific research. These are traditions which history describes under such rubrics as "Ptolemaic Astronomy" (or "Copernican"), "Aristotelian dynamics" (or "Newtonian"), "corpuscular optics" (or "wave optics"), and so on.

Normal Science

A term closely related to paradigm in Kuhn's scheme is "normal science". Normal science is what scientists do when they work routinely within established doctrinaire paradigms.

It is the science we find in textbooks.

Scientists use paradigms as resources to refine theories, explain puzzling data, establish increasingly precise measures of standards, and do other necessary work to expand the boundaries of normal science.

Revolutionary Science

The serene stability of normal science is occasionally punctuated by irresolvable crisis. A point is reached when the crisis can only be solved by *revolution*. "Revolutionary science" takes over and old paradigms give way to new ones. But what was once revolutionary itself becomes the new orthodoxy. And the cycle begins again.

> Science advances through cycles of normal science followed by revolutionary science.

> Each paradigm produces a major work that defines and shapes it.

Aristotle's *Physics*, Newton's *Principia* and *Opticks*, and Lyell's *Geology* are examples of works that defined the paradigms of particular branches of science at particular times.

In sharp contrast to the traditional picture of science as a progressive, gradual and cumulative acquisition of knowledge, based on rationally chosen experimental frameworks, Kuhn presented "normal" science as a dogmatic enterprise.

The Enemy of Science

Not surprisingly, *Structure* generated a great deal of controversy. Scientists were repelled by its suggestion that far from being heroic, open-minded, disinterested seekers of Truth and interrogators of nature and reality, they were a specialized priesthood promoting their own specific denominational theologies. Philosophers of science also found Kuhn's relativism quite repugnant.

Popper was amongst the first persons to recognize the importance of Kuhn – in <u>Structure</u>, he saw a threat to the future of science.

Kuhn's idea of "normal science" is an enemy of science and civilization.

In Opposition to Kuhn

In July 1965, Popper and his group organized an International Colloquium in the Philosophy of Science with the explicit aim of destroying Kuhn. The idea of the Colloquium, backed by a whole range of institutions – including the British Society for the Philosophy of Science, London School of Economics and International Union of History and Philosophy of Science – was to pit Kuhn against the combined might of the British philosophers of science.

LADEEZ & GENNL'MEN!!!
ONLY TONIGHT TWO
FOR ONE NIGHT ONLY
BEING FOR THE BENEFIT OF MR KITE
AN INTERNATIONAL COLLOQUIUM
In The Philosophy Of Science
WITHOUT THE AID OF A NET!
A Heavyweight Bout
3 Falls, 3 Submissions or a Knockout
TO DECIDE THE WINNER:

Karl Popper
The Masked Marauder
✻ VS. ✻

Thomas Kuhn
The Anarcho-Scientist From Hell

followed by Ventriloquist and Drag Artiste

I survived.

The result of the debates, including Kuhn's replies, were published as <u>Criticism and the Growth of Knowledge</u> *(1970).*

The End of "Dominant Notions"

By the early 1970s, *Structure* was accepted as a truly revolutionary work. According to Ian Hacking, *Structure* spelled the end of the following notions …

Realism: that science is an attempt to find out about one real world; that truths about the world are true regardless of what people think; that the truth of science reflects some aspect of reality.

Demarcation: that there is a sharp distinction between scientific theories and other kinds of belief systems.

Cumulation: that science is cumulative and builds on what is already known – for instance, Einstein being a generalization of Newton.

Observer–theory distinction: that there is a fairly sharp contrast between reports of observation and statements of theory.

Foundations: that observation and experiment provide the foundations for and justification of hypotheses and theories.

Deductive structure of theories: that tests of theories proceed by deducing observation-reports from theoretical postulates.

Precision: that scientific concepts are rather precise and the terms used in science have fixed meanings.

Discovery and justification: that there are separate contexts of discovery and justification; that we should distinguish the psychological or social circumstances in which a discovery is made from the logical basis for justifying belief in facts that have been discovered.

The unity of science: that there should be one science about the one real world; less profound sciences are reducible to more profound ones – psychology is reducible to biology, biology to chemistry, chemistry to physics.

Isn't that rather a lot of "stuff" on one page?

Stuffed to the gunwhales, I'd say... More tea?

Is Kuhn a Radical?

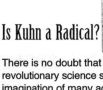

There is no doubt that Kuhn's talk of revolutionary science sparked the imagination of many academic radicals in the 1960s and 70s. However, it would be misguided to see Kuhn himself as a radical, or *Structure* as a work of great radical thought.

> Kuhn should really be understood as part of the conservative élitist tradition in scientific and political thought. We should see him operating at two historical levels.

1. The first level concerns the need to protect both the autonomy and authority of science in a political crisis period – the Cold War – that witnessed increasing suspicion of science and greater calls for its social control.

2. The second level makes Kuhn part of a larger tradition of conservative political thought, going back to **Plato** (c. 428–347 BC), which distrusts public involvement in determining the truths by which society should live.

The Birth of "BIG SCIENCE"

The most immediate level is the Cold War context from which Kuhn's account of science emerged. Kuhn trained at Harvard as a physicist in order to follow up the great problems of natural philosophy pursued by Newton and **Albert Einstein** (1879–1955).

But my first taste of life as a physicist was jamming German radar signals in World War II ...

> This experience, combined with the explosion of the first atomic bombs, marked the beginning of "Big Science".

> "Big Science" meant that scientific research would be driven by technology – both in terms of the constitution of its research agenda and its applications in the larger society.

Kuhn was rescued from complete disillusionment with physics by **James Bryant Conant** (1893–1978), President of Harvard University and chief scientific administrator of the US atomic-bomb project. Kuhn regarded Conant as the smartest man he had ever met.

Conant found a place for Kuhn in the General Education in Science programme, designed to make America's future leaders sympathetic to scientific research.

Supporting Big Science

Conant's idea was to have students see the "Big Science" projects of their own day through the ideals informing the "Little Science" projects that had enabled the natural sciences in the modern era to be part of the West's cultural inheritance.

> The value of a particle accelerator must not be judged by its cost or potential contribution to nuclear energy, but by the <u>theoretical principles</u> it enables one to test ...

> In other words, a continuation of the "classical quest" for a unified account of physical reality.

> By focusing students' minds in this way, future decision-makers would continue to support science without imposing too many external constraints.

However, Kuhn did not realize that an account of science which did not highlight its social, economic or technological impacts would be readily appropriated by non-natural scientists for their own purposes – including Science Studies practitioners! Kuhn's model of scientific change unwittingly empowered a vast range of inquirers that neither Conant nor Kuhn had intended.

Feyerabend, the Anarchist

Paul Feyerabend (1924–94) was one of the earliest, persistent and influential critics of the positivist interpretation of science. Although his criticism of science is somewhat similar to that of Kuhn, his views are much more radical. Born in Austria, Feyerabend had a varied career …

... a spell in the army and with the Communist playwright <u>Bert Brecht</u> (1898–1956) before I became a philosopher of science.

He debated brilliantly on behalf of Popper. By the time he participated in the famous Colloquium against Kuhn, organized by Popper and his group, he had already developed drastically different ideas about science.

Anything Goes

Feyerabend's most central idea was "epistemological anarchism". In *Against Method* (1975), he argued that any principle of Scientific Method has been violated by some great scientist – Galileo is one example amongst many others. So, if there is a Scientific Method at all, it can only be – "anything goes".

> Science is an essentially anarchistic enterprise. Theoretical anarchism is more humanitarian and more likely to encourage progress than its law-and-order alternatives.

Feyerabend demonstrates by an examination of historical episodes and by analysis of the relation between idea and action. The only principle that does not inhibit progress is – *anything goes*.

A Free-for-all

For Feyerabend, science has no claims to superiority over other systems of thought such as religion and magic. As a tactical anarchist, he held classes at the University of Berkeley where he famously invited creationists, Darwinists, witches and other "truth peddlers" to defend their opinions in front of the students.

In *Farewell to Reason* (1987), Feyerabend attacked the very idea of scientific rationalism.

Most great scientists and philosophers – from Galileo to Popper – are irrational dogmatists.

Science's appeal to reason is nothing but empty and tyrannical.

Science must become subordinate to the needs of citizens and communities.

Sociology of Scientific Knowledge

The Sociology of Scientific Knowledge (SSK) is based on the assumption that our natural reasoning capacity and sense perceptions are not sufficient conditions for the production of scientific knowledge.

Sociologists studying science look at contents, style, methods, conventions and institutions for the answers.

What else is needed?

Originally, science was actually excluded from sociology of knowledge.

Karl Mannheim (1893–1947), the founding father, believed that scientific knowledge was universal – its objectivity transcended specific cultural origins – and hence science was beyond sociological inquiry.

The Spirit of Science

Several types of sociology of science were developed within these limits after the Second World War. The most influential was that proposed by the American sociologist **R.K. Merton** (b.1910) who systematized the normative pronouncements of famous scientists.

> The scientific movement of the 17th century was a result not so much of the prevailing socio-economic conditions but more a product of the Protestant ethic.

> An idea he borrowed from me – Max Weber (1864–1920) – and my book *The Protestant Ethic and the Spirit of Capitalism* (1904–5).

In the late 1960s, Mannheim's strictures were unceremoniously ejected by the "Strong Programme".

The Strong Programme

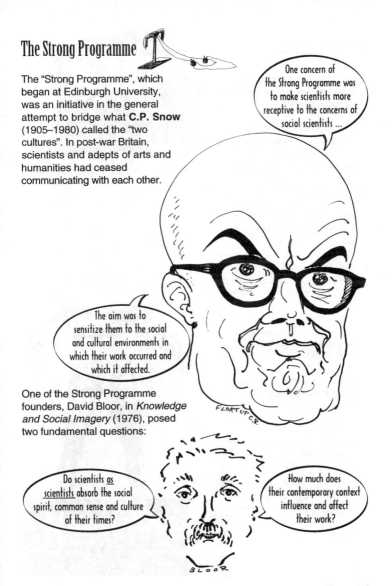

The "Strong Programme", which began at Edinburgh University, was an initiative in the general attempt to bridge what **C.P. Snow** (1905–1980) called the "two cultures". In post-war Britain, scientists and adepts of arts and humanities had ceased communicating with each other.

One concern of the Strong Programme was to make scientists more receptive to the concerns of social scientists ...

The aim was to sensitize them to the social and cultural environments in which their work occurred and which it affected.

One of the Strong Programme founders, David Bloor, in *Knowledge and Social Imagery* (1976), posed two fundamental questions:

Do scientists <u>as scientists</u> absorb the social spirit, common sense and culture of their times?

How much does their contemporary context influence and affect their work?

The Basics of SSK

The proponents of the Strong Programme argue that SSK has four basic elements.

1. SSK discovers the conditions – economic, political, social, as well as psychological – that bring about states of knowledge.

2. SSK is impartial in its selection of what is studied. It gives equal emphasis to true and false knowledge, successes and failures of science.

3. SSK is consistent (or uses "symmetry") in its explanation of selected instances of scientific knowledge. It would not, for example, explain a "false" belief with sociological cause or use a rationalist cause to explain a "true" belief.

4. The models of explanation of SSK are applicable to sociology itself.

In its early phases, the Strong Programme was seen as truly radical and subverting of science.

Science as Social Construction

Certain sociologists of science argue that science is *socially constructed* and not determined by the world or some "physical reality" out there. These scholars are called "Constructionists". Constructionists study specific historic or contemporary episodes in science. They also carry out "field research" in laboratories.

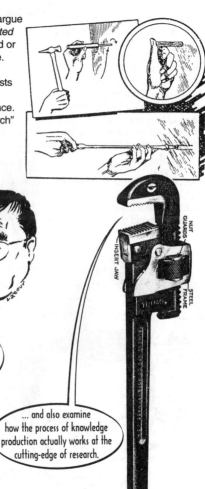

> We interrogate the "facts" of science and the "truths" they are supposed to express ...

> ... and also examine how the process of knowledge production actually works at the cutting-edge of research.

The Effect of Reality

The most famous constructionist study is *Laboratory Life: Social Construction of Scientific Facts* (1979; 1986) in which Bruno Latour and Steve Woolgar examined the detailed history of a single fact: the existence of Thyrotropin Releasing Factor (Hormone), or TRF(H) for short. Latour and Woolgar show that TRF(H) has meaning and significance according to the *context* in which it is used.

It has a different significance for each group of specialists – medical doctors, endocrinologists, researchers and graduate students who use it as a tool in setting up bioassays.

> For specialists who have spent their entire professional career studying it, TRF(H) represents a subfield.

> But outside this network, TRF(H) does not exist.

Latour and Woolgar also suggest that the transformation of statement into fact is reversible: that is, reality can also be deconstructed. Reality cannot be used to explain why a statement becomes a fact, since it is only after a fact has been constructed that the *effect of reality* is obtained.

The Construction of Objectivity

Before Latour and Woolgar's investigation, Ian Mitroff's *The Subjective Side of Science* (1974) examined the perceptions, cherished theories and published results of scientists who analysed lunar rocks brought back by Apollo 11.

In almost all cases, these scientists found what they expected to find.

Er... It's a lump of rock, isn't it?

Mitroff reluctantly concluded: scientific objectivity is nothing but a socially constructed charade.

The Science Tribe

In her seminal work, *The Manufacture of Knowledge* (1981), Karin Knorr-Cetina studied scientists in a laboratory like a tribe in the jungle.

undergrowth. When I observed the process of knowledge production in the laboratory I found an opportunistic logic at work. Scientific Method turns out to be a locally situated, locally proliferated form of practice, rather than a paradigm of non-local universality. It is context-impregnated, rather than context-free. And it can be seen as rooted in the site of local social action, just as other forms of life are.

Thursday 25th July.
The sweat is dripping from the ceilings in this tropic...

All this raises the question: can there ever be such things as value-neutral "objective facts"?

Theory-laden Observations

The constructionists instead take the view that scientists do not make observations in isolation but within a well-defined theory. These observations – and the data collection that goes with them – are designed either to refute a theory or provide support for it.

> And, as Kuhn has shown, theories exist within <u>paradigms</u>. Observations themselves have validity only within those specific theories.

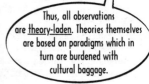

> Thus, all observations are <u>theory-laden</u>. Theories themselves are based on paradigms which in turn are burdened with cultural baggage.

The Context of "Tradition"

Defenders of the Strong Programme argue that it is not so much the observations in science that are "theory-laden" but rather the reports of the observations. How an observation is reported depends on the *tradition* within which a scientist is working. The interpretation of an observation involves bringing to bear the resources of a tradition.

According to the Edinburgh school, theories are not fixed in time. Nor can they be identified with a set of fixed statements.

Two scientists working in different traditions may observe the same thing, but report and interpret the same results in different ways.

Association of theories with the names of Great Scientists – "Newton's Theory", "Einstein's Theory" – creates this illusion. It is better to think of scientific theories as *evolving institutions*. A detailed examination of "Mendel's theory" shows us how many twists and turns it has taken since Mendel first formulated it.

Feminist Criticism

Feminist scholarship of science developed in parallel with SSK, as well as with the radical criticism of science outside academia. It has shown that the focus on quantitative measure, analysis of variation and impersonal, excessively abstract, conceptual schemes is both a distinctively masculine tendency and also one that serves to hide its own gendered character.

The prioritizing of mathematics and abstract thought, standards of objectivity, the construction of scientific method and the instrumental nature of scientific rationality are all based on the notion of <u>ideal masculinity</u>.

Feminist criticism first began with the exploration of issues relating to women's participation in science.

Women in Science

Science has systematically marginalized and undervalued women's contributions. Gender stereotyping actually begins in the cradle and accumulates through childhood, adolescence and adulthood to discourage women and encourage men to adopt those kinds of thinking and motor activities necessary for skills in scientific, mathematical and engineering work.

Not surprisingly, less than a quarter of US scientists are women.

"Women's struggle to break into science can be seen as a parallel with their struggle to break into the clergy. Christians traditionally believed that God had written two books – Scripture and Nature, both being an expression of the Divine Word. For much of the past two thousand years, the study of Scripture was seen as a task fitting to men alone. So too the study of Nature – God's 'other Book' – was long seen as an essentially male activity. Just as women had to fight for the right to be theologians and priests, so too they have had to battle the 'Church of Science' for the right to be scientists."
Margaret Wertheim, author of *Pythagoras' Trousers* (1995)

What was it again? "The blokes shall inherit the Earth"?

The Segregation of Women in Science

Women began to choose science as a career in the period between 1820 and 1920. This era saw a thousand-fold increase in the participation of women in science in the US. But the growth occurred at a price.

Women had to accept a pattern of segregated employment and under-recognition.

Try as we might, most women could not escape that.

O.K., O.K.... Forget your PhD and put the kettle on, will you?

By 1920, this pattern was well established. Despite much protest by feminists, women's subsequent experience in science was one of containment within demarcated limits. They were confined to fields such as "home economics" and "cosmetic chemistry". Expansion into new areas was limited.

The Invisible Woman in the Lab

Nowadays, most women scientists are primarily to be found in the lower echelons of the scientific enterprise, doing rank-and-file work in laboratories. Women scientists running their own laboratories are extremely rare and few can find the resources to carry out independent research. In most cases, their work is systematically undervalued, relative to similar achievements by men.

A number of studies have shown that scientific work done by women is invisible to men ...

... even when it is <u>objectively indistinguishable</u> from men's work.

Ejecting women in the name of higher standards was one way of keeping women away from science.

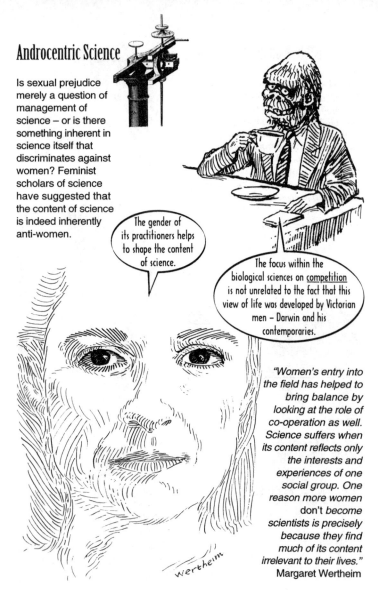

Sandra Harding, Professor of Philosophy at the University of Delaware and author of the influential *The Science Question in Feminism*, offers an insight into how science is saturated with "androcentric" imprints. Consider, for example, traditional evolutionary theories that explain the roots of present human behaviour. The origins of Western, middle-class social life – where men go out to do what a man's got to do and kitchen-bound women tend the babies – are to be found in the bonding of "man-the-hunter".

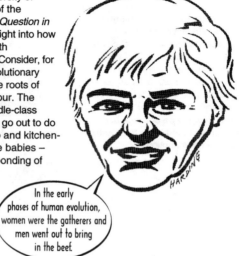

In the early phases of human evolution, women were the gatherers and men went out to bring in the beef.

This theory is based on the discovery of chipped stones that are said to provide evidence for the male invention of tools for use in the hunting and preparation of game.

Turn the page for an alternative view.

Women as Providers

But you can look at the same stones with different cultural perceptions. We know that cultures exist in the present day in which women are the main providers of the group. You can then argue that these stones were used by women to kill animals, cut meat, dig up roots, break down seed pods, or hammer and soften tough roots to prepare them for consumption.

You now have a totally different hypothesis ...

And the course of the whole evolutionary theory changes!

Other developments in science – such as the rise of IQ tests, behavioural conditioning, foetal research and socio-biology – can also be analysed with similar logic.

More Women in Science

Would a fair representation of women in science change anything? To begin with, it would have obvious economic advantages.

Knowledge-based economies, in dire need of trained scientists, cannot afford to squander half of their scientific potential.

More women in science would also open up science to a wider range of material and social problems.

For example, the problems of the Third World would receive greater emphasis and more research support.

But the feminist critique goes much deeper ...

Strong Objectivity

Sandra Harding suggests that women would introduce a shift away from conventional scientific methods of objectivity to what she calls "strong objectivity".

Strong objectivity requires that scientists take account of the perspectives of "outsiders" in their descriptions and explanations of the subject of scientific inquiry...

social scientists

environmentalists

housewives

non-Western cultures

Strong objectivity leads to standpoint epistemologies that use fundamental questions raised by groups marginalized from institutionalized power to shape inquiry and knowledge production.

Feminist analysis is not culturally neutral elaborations of people's social experience – of what members of marginalized groups say about their lives – but theoretical reflections on them. Marginalized experiences, and what marginalized peoples say, are crucial guides to the new questions that can be asked about nature, science and social relations.

> Such questions arise out of the gap between marginalized interests and consciousness ...

> ... and the way the dominant conceptual schemes organize social relations, including those of scientific and technological change.

Standpoint epistemologies propose that scrutiny of institutionalized power-imbalances begins with marginalized lives. This gives a critical edge for formulating new questions. Everyone's knowledge about institutionalized power and its effects is thus expanded. Feminist science and technology studies have undertaken just such projects.

Responsible Rationality

In a similar vein, Hilary Rose, doyenne of British Science Studies and author of *Love, Knowledge, Power* (1994), has developed the idea of "responsible rationality" that restores care and concern within scientific objectivity.

> Under the banner of "objectivity" and "rationality", the life sciences have been constructing <u>difference</u> as both natural and hierarchical. It was important for us feminists to challenge this.

Reproductive Labour

During the radical 1960s and 70s, when culture was preoccupied with production, a central feminist project was to foreground human *reproduction*. There were celebratory essentialist versions and Marxist feminist versions which rooted gender difference in the division of *reproductive labour*. Both essentialists and Marxist feminists shared a bio-social view that dependent human beings – especially small children – needed love or caring rationality to survive.

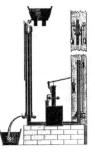

With this doubleness, the ability of the old construction of scientific rationality to exclude love and responsibility is weakened.

The concept of a baby was thus neither a biological category nor a social category, but both.

Such weakening is crucial if the techno-sciences set to dominate the 21st century are to be reshaped to enable the survival of both "society" and "science". Environmentalists in their concern to defend the socio-ecological system have come to a remarkably similar position.

Post-colonial Science Criticism

Like feminist scholars, post-colonial critics argue that real change can come about only through a fundamental transformation of concepts, methods and interpretations in science – a complete re-orientation in the logic of scientific discovery.

With the sole exception of feminist scholars, post-colonial criticism was mostly ignored by mainstream Science Studies.

Only after the 1990s, when the sheer qualitative and quantitative output was almost impossible to render invisible, did post-colonial Science Studies begin to make an impact on Western Science Studies.

Science and Empire

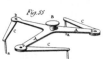

Post-colonial scholarship of science seeks to establish the connection between colonialism – including neo-colonialism – and the progress of Western science. For example, in his several books, Deepak Kumar, the Indian historian and philosopher of science, has sought to demonstrate that British colonialism in India played a major part in how European science developed.

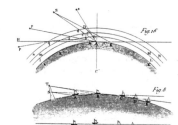

> The British needed better navigation, so they built observatories, funded astronomers and kept systematic records of their voyages.

> The first European sciences to be established in India were, not surprisingly, geography and botany.

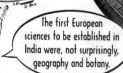

Throughout the Raj, British science progressed primarily because of military, economic and political demands of the British, and not because of the purported greater rationality of science or the alleged commitment of scientists to the pursuit of disinterested truths.

Consider the motto of Imperial College, London:

> Science is the pride and shield of Empire.

It was the sword as well.

Science and empire developed and grew together, each enhancing and sustaining the other. Indeed, we can trace the establishment of many institutions of science to the period when Europe began its imperial adventure. Schools of Tropical Medicine in London and Liverpool were established in 1899 with the sole aim of aiding empire builders.

The study of "tropical diseases" did not embrace all tropical diseases but only those relevant to British interests.

Tropical medicine concentrated on the tropical ailments of Europeans.

Only when it was extended to natives in 1918 was the discovery of endemic disease and malnutrition made. Tropical crops were almost always cash crops.

Imperial Geography

The political ambitions of East India Company necessitated a thorough geographical knowledge – hence the Geological Survey of India which got the maximum patronage of the British government. When completed in 1856, it was described as representing "the common sense of the Empire" and was used to justify the colonization of India.

Half the survey was devoted to the study of coal mines – because that's what the British were interested in.

In Egypt and the Sudan, the British overlooked schistosomiasis for decades – now it is recognized as a major endemic disease in these regions.

There was no scientific education in the colonies till 1940. Natives, assumed to be backward in nature, worked as technicians and laboratory assistants, but never qualified as doctors or scientists or researchers.

What Happened Under Colonialism?

Science adopted specific policies towards non-Western sciences during colonialism. Western scientists assumed that no other sciences could generate the laws of gravity or antibiotics and only Western science could discover all the laws of nature. A policy of ruthlessly suppressing non-Western and indigenous sciences was thus pursued.

Specifically, Western science appropriated and integrated non-Western science without acknowledgement. The pre-Colombian agriculture that provided potatoes for almost every European ecological niche became part of European science. Mathematical and astronomical achievements from Arabic and Indian cultures provide another example. Islamic medicine was almost totally appropriated. The magnetic needle, the rudder, gunpowder and many other technologies useful to European sciences were borrowed from China. Knowledge of local geographies, geologies, animals, plants, classification schemes, medicines, pharmacologies, agriculture and navigational techniques was provided by the knowledge traditions of non-Europeans. After appropriating and plagiarizing non-Western knowledge, Western science recycled it as its own.

Non-Western sciences were made invisible – by writing them out of history. This occurred during the Enlightenment period, when, for example, the French *philosophes* produced their great encyclopaedia. The period that fell between ancient Classical times and the Renaissance then came to be named the "Dark Ages" when simply nothing happened.

Western prejudice denigrated, abused and then ruthlessly suppressed non-Western science. In the colonies, anything to do with indigenous science and learning was made illegal. In Algeria and Tunisia, for example, the French made the practice of Islamic medicine a crime punishable by death. Indeed, countless Islamic doctors were executed. In Indonesia, the Dutch closed all universities and institutions of higher learning and made it illegal for the natives to be educated.

Empirical History of Islamic Science

Post-colonial Science Studies began with empirical work on the history of Islamic, Indian and Chinese civilizations. During the 1960s and 70s, original work in the history of Islamic science revealed how truly awesome – both in depth and breadth – were the scientific achievements of Muslim civilization. An inkling of that was already provided by George Sarton in his *Introduction to the History of Science* (1927).

> But the history of Islamic science really came into its own with Fuat Sezgin's monumental work on Islamic science, <u>Gesichte des Arabischen Schrifttums</u> (numerous volumes, 1967–) ...

> ... and the efforts of scholars in France working with Roshdi Rashed.

Since then, the output of numerous scholars, including the work of Turkish scholar Ekmeleddin Ihsanoglu on Ottoman science, has established that science as we know it today would have been inconceivable without Islamic science.

Indian and Chinese Science

The history of Indian science experienced a similar revival with the publication of the bibliographic work of A. Rahman and *A Concise History of Science in India* (two volumes), edited by D.M. Bose, S.N. Sen and P.V. Sharma.

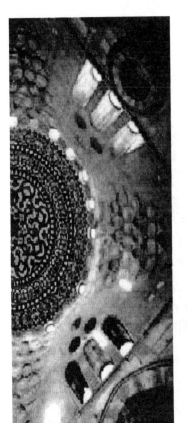

A similar boost was given to the history of Chinese science by Joseph Needham's *Science and Civilisation in China* (seven volumes, 1954–), which was later built upon by indigenous works such as Peng Yoke Ho's *Li, Qi and Shu: An Introduction to Science and Civilization in China* (1985).

Rediscovery of Civilizational Science

Finally, post-colonial scholarship of science seeks to re-establish the practice of Islamic, Indian or Chinese science in *contemporary* times. There is, for example, a whole discourse of contemporary Islamic science devoted to exploring how a science based on the Islamic notions of nature, unity of knowledge and values, public interest and so on, could be shaped.

Blimey! Gold already!

In my book, <u>The Touch of Midas</u> (1984), a contemporary notion of Islamic science was developed for the first time. It was later elaborated in <u>Explorations in Islamic Science</u> (1989).

Framework for Islamic Science

The contemporary reformulation of Islamic science is based on a conceptual matrix derived from the Qur'an. These concepts generate the basic values of Islamic scientific culture and form a parameter within which science advances. There are ten such concepts, four standing alone and three opposing pairs ...

tawheed (unity)
|
khalifah (trusteeship)
|
ibadah (worship)
|
ilm (knowledge)

Positive — **halal** (praiseworthy) -- **haram** (blameworthy) — Negative
adl (social justice) -- **zulm** (tyranny)
istislah (public interest) -- **dhiya** (waste)

> When translated into values, this system of concepts embraces the nature of scientific inquiry in its totality.

> It integrates facts and values and institutionalizes a system of knowing that is based on accountability and social responsibility.

AVERRÖES - MUSLIM PHILOSOPHER

Tawheed and *Khalifah*

How do these values shape scientific and technological activity?

Usually, the concept of *tawheed* is translated as "unity of God". It becomes an all-embracing value when this unity is asserted in the unity of humanity, unity of person and nature, and the unity of knowledge and values.

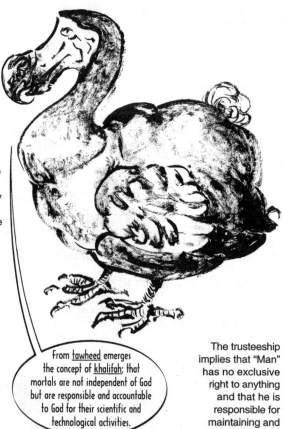

From <u>tawheed</u> emerges the concept of <u>khalifah:</u> that mortals are not independent of God but are responsible and accountable to God for their scientific and technological activities.

The trusteeship implies that "Man" has no exclusive right to anything and that he is responsible for maintaining and preserving the integrity of the abode of his terrestrial journey.

Ibadah: Non-violent Contemplation

But just because knowledge cannot be sought for the outright exploitation of nature, one is not reduced to being a passive observer. On the contrary, contemplation (*ibadah*) is an obligation, for it leads to an awareness of *tawheed* and *khalifah*. It is this contemplation that serves as an integrating factor for scientific activity and a system of Islamic values. *Ibadah*, or the contemplation of the unity of God, has many manifestations, of which the pursuit of knowledge is the major one.

If scientific enterprise is an act of contemplation – a form of worship – it goes without saying that it cannot involve any acts of violence towards nature or creation.

Indeed, it cannot lead to waste (<u>dhiya</u>), or any form of violence, oppression or tyranny (<u>zulm</u>) ...

Or be pursued for unworthy goals (<u>haram</u>).

It can only be based on praiseworthy goals (<u>halal</u>) on behalf of public good (<u>istislah</u>) and overall promotion of social, economic and cultural justice (<u>adl</u>).

Such a framework propelled Islamic science in history towards its zenith without restricting freedom of inquiry or producing adverse effects on society. The contemporary research on rediscovering the nature and style of Islamic science would have tremendous effect both on policies and the content of science in the Muslim world.

Rediscovering Indian Science

A similar discourse on Indian science emerged during the 1980s and 90s. It is most strongly associated with the numerous academic and radical groups involved in the periodic organization of the Congress on Traditional Sciences and Technologies of India.

If houses can be built only with cement and steel, then there may be no way we can think of to provide housing for all.

The picture changes substantially if we include the wide variety of materials and techniques traditionally employed by our people.

If we also include the wide variety of proven medicine, practices and principles that have been indigenously involved in healthcare in our society, then the resource position on the healthcare front may not appear as bleak as it now seems.

Walking on Two Legs

> If the wide range of materials and techniques that our farmers have traditionally employed to ensure land fertility, pest control, high yield etc. are included in the list of resources at our command ...

> ... then the prospect of enhancing food production substantially in an ecologically and economically sound manner may not appear as daunting as it seems to be now.

Centre for Indian Knowledge Systems
Real world solutions from traditional Indian sciences

> We recognize a wide variety of skills and knowledge that the Indian people possess, which, if properly understood and recognized, can make a substantial contribution to productive efforts and endeavours.

India has laboured under the severe yoke of "resource scarcity" largely because it did not recognize the existence of an indigenous, traditional resource base. "Resources" only included those materials, processes, skills and theories that the West had been using after achieving full modernization and international domination. Limiting India to these options alone was almost like entering a race with both feet tightly tied together.

> Ready...
> Steady...
> Go!

The Western View of Nature

The main post-colonial criticism of science concerns its basic assumptions about nature, universe, time and logic. All these assumptions – as post-colonial critics such as Indian scholars Ashis Nandy and Claude Alvares argue – are *ethnocentric*.

In modern Western science, nature is seen as hostile, something to be dominated. The Western "disenchantment of nature" was a crucial element in the shift from the medieval to the modern mentality, from feudalism to capitalism, from Ptolemaic to Galilean astronomy, and from Aristotelian to Newtonian physics.

In this picture, Man stands apart from nature, on a higher level, ready to subjugate her.

Nature yields her secrets under "torture".

Assumptions Shape Science

Similarly, while modern science sees time as linear, other cultures view it as cyclical, as in Hinduism, or as a tapestry weaving the present with eternal time in the Hereafter, as in Islam.

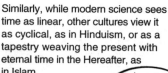

Modern science operates on the basis of "either/or" Aristotelian logic ...

X is either A or non-A.

In Hinduism, logic can be four-fold or even seven-fold ...

X is neither A, nor non-A, nor both A and non-A, nor neither A nor non-A.

The four-fold Hindu logic is both symbolic as well as a logic of cognition and can achieve precise, unambiguous formulation of universal statements without quantification.

The metaphysical assumptions of modern science make it specifically Western in its main characteristics.

What is Assumed "Efficient"?

These metaphysical assumptions of Western science are reflected in its contents. Certain laws of science, as Indian physicists have begun to demonstrate, are formulated in an ethnocentric and racist way. The Second Law of Thermodynamics, so central to classical physics, is a case in point.

> Due to its industrial origins, the Second Law presents a definition of efficiency that favours high temperatures and the allocation of resources to big industry.

> Work done at ordinary temperatures is by definition inefficient.

Both nature and the non-Western world become losers in this new definition. For example, the monsoon – transporting millions of tons of water across a subcontinent – is "inefficient" since it does its work at ordinary temperatures. Similarly, traditional crafts and technologies are designated as inefficient and marginalized.

Assumptions of Genetic Differences

In biology, social Darwinism is a direct product of the laws of evolutionary theories. Genetic research appears to be obsessed with how variations in genes account for differences among people. Although we share between 99.7 and 99.9 per cent of our genes with everyone, genetic research has been targeted towards the minute percentage of genes that are different in order to discover correlations between racial characteristics such as skin colour, and either intelligence or "troublesome" behaviour.

Enlightened social pressures often push the racist elements of science to the sidelines.

Trust me. I'm a scientist. No, really... Trust me.

But the inherent metaphysics of genetics ensures that they reappear in new disguise.

Witness how *eugenics* keeps reappearing with persistent regularity. The institution of IQ tests, behavioural conditioning, foetal research and socio-biology are all indications of the racial bias inherent in modern science.

The "Value" of Science

Science in developing countries has persistently reflected the priorities of the West.

The needs and requirements of middle-class Western society are emphasized.

Rather than the needs, requirements and conditions of our own society.

In over five decades of science development, most of the Third World countries have nothing to show for it. The benefits of science just refuse to trickle down to the poor.

The Myth of Neutrality

Even if we were to ignore all other arguments and evidence, the very claim of modern science to be value-free and neutral would itself mark modern science as ethnocentric and a distinctively Western enterprise.

Both claiming and maximizing cultural neutrality is itself a specific Western cultural value.

Non-Western cultures do not value neutrality for its own sake but emphasize and encourage the connection between knowledge and values.

By deliberately trying to hide its values under the carpet, by pretending to be neutral, by attempting to monopolize the notion of absolute truth, Western science has transformed itself into a dominant and dominating ideology.

The inherent biases of science are scrutinized by an academic movement called *social epistemology*.

Social Epistemology

Social epistemology emerged in the 1980s as a critical movement concerned with the fundamental questions about the nature of knowledge. Steve Fuller, the founder of the school of social epistemology, and his students, were concerned with attempts to reconcile *normative* and *empirical* approaches to the study of science.

For social epistemology, science is the systematic pursuit of knowledge, whether it be of the natural or the human world.

Normative approaches have been traditionally championed by philosophers who are concerned with how science "ought to be" ...

What Social Epistemology Asks ...

What sort of knowledge do we want?
For what ends?
Who should be producing it?
On behalf of whom?
How should we be using it?

Science Communication

Another way of pursuing social epistemology is by promoting the importance of rhetoric in the curriculum, specifically by encouraging specialists in Science Studies to join "science communication" programmes in which people who already hold science degrees seek to become part of the "public relations" arm of science.

> Traditionally, these programmes have been devoted to revealing all the benefits of science, while hiding the costs.

> However, in today's more sceptical climate, they have become vehicles for renegotiating science's social contract with the public.

> Now that we've been reassured about our local radioactivity leak, I feel great!

> The pluses and minuses of scientific research are argued in ways that encourage the public to ask "what's in it for them".

Multi-culturalism and Scientific Knowledges

Social epistemology has been instrumental in promoting multi-culturalism as a vehicle for envisioning alternative ends and means of organizing the production of knowledge. However, the aim here is less on preserving distinct "local knowledges", such as in museum exhibits, than in enabling one culture to learn from the successes and failures of other cultures' knowledge-producing practices.

For example, Islam has much to teach the West about how a holistic form of inquiry might be pursued.

Japan provides an example in the other direction, namely, knowledge as purely instrumental.

In neither case is knowledge simply pursued "for its own sake", as the Western epistemologies would so often have it.

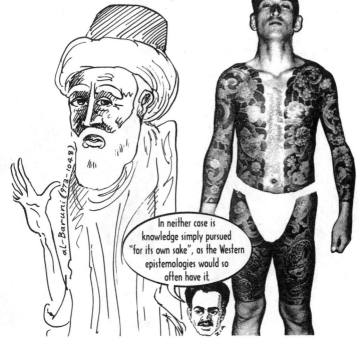

al-Baruni (973-1048)

Science Wars

For much of the second half of the 20th century, scientists took the criticism of sociologists of science, social constructionists, social epistemologists, feminists and post-colonial scholars with – shall we say – some grace. They continued to do what they always did, with an occasional senior statesman of science – usually Steven Weinberg – standing up to defend the good ol' values of science.

But in the 1990s, public disenchantment with science reached an all time high ...

Animal rights activists started picketing laboratories.

Funding for big science, such as super-conducting super-collider projects, began to be squeezed. A full onslaught against the "science critics" was launched.

In Defence of Science

A broad coalition of scientists, social scientists and other scholars was mobilized for the defence of science through a series of lavish, well-funded and highly publicized conferences. The most effective of these was the *Flight from Science and Reason* conference, sponsored by the New York Academy of Science, held in New York during the summer of 1995.

Science is under serious threat from sociologists, historians, philosophers and feminists who work in the field of "Science and Technology Studies" (STS).

We attack the "social theories" of science and declare feminist epistemology a "dead horse".

The criticism of science is a "common nonsense" and most critics of science are "charlatans".

The issues, the conference declared, are those of Reason and its application in science – and the status of these in our time.

Against the "Academic Left"

Defenders of the purity of science were convinced that there was a conspiracy from the "academic left" against science.

The academic left – a large and influential segment of the American academic community – dislikes science.

The much-cited work by Paul Gross and Norman Levitt, *Higher Superstition: The Academic Left and Its Quarrels with Science* (1994), became an unofficial manifesto of the defenders of science.

Hostility extends to the social structures through which science is institutionalized...

...to the system of education by which professional scientists are produced...

...and to a mentality that is taken, rightly or wrongly, as characteristic of scientists.

There is open hostility from the "left" towards the actual content of scientific knowledge and towards the assumption – which one might have supposed universal among educated people – that scientific knowledge is reasonably reliable and rests on a sound methodology.

The medieval hostility of the critics of science is a clear rejection of the strongest heritage of the Enlightenment and a denial of progress.

Clearly, scientists were ready to take their attack deep into enemy territory.

Enter, Sokal *(stage right)*

The Duke University journal *Social Text* is perhaps one of the most sacred precincts of the Cultural Studies brigade. On the cover of the Spring/Summer 1996 issue, we read:

Science Wars. As part of the campaign against "political correctness" the history and theory of Science Studies is increasingly subject to intense political scrutiny. In this special issue edited by Andrew Ross, many of the leading figures in the social and cultural study of science respond to recent debates in the field. Contributors to this issue include Sandra Harding, Steve Fuller, Emily Martin, Hilary Rose, Langdon Winner, Dorothy Nelkin, George Levine, Sharon Traweek, Sarah Franklin, Richard Lewin, Joel Kovel, Stanley Aronowitz, Andrew Ross, Les Levidow, and Alan Sokal.

Hi! Watch this space!

SOKAL

46-47

The journal editor, Andrew Ross, describes science as a new religion and dismisses *Higher Superstition* as a shallow "shrill" work belonging to the well-established right-wing scholarly tradition of "crying wolf".

After the orthodox pronouncements from the Curia – the college of cardinals of Science Studies – comes the curious contribution by Alan Sokal, a Professor of Physics at New York University, entitled "Transgressing the Boundaries: Towards a Transformative Hermeneutics of Quantum Gravity". The paper is a trifle unusual – even by the constructionist tradition of extreme relativism.

> It suggests that π (pi), far from being a constant and universal, is actually relative to the position of an observer and is thus subject to "ineluctable historicity".

The bibliography clearly reads like a deliberately constructed "Who's Who" of science critics and bears little relationship to the contents of the paper. And it contains embarrassingly flattering citations from the works of Andrew Ross and Stanley Aronowitz, editors of the journal. Yet, the editors of *Social Text* themselves failed to grasp its significance.

> It was a hoax.

Blitzkrieg on Postmodernism

When Sokal revealed his hoax, "Science Wars" went public in a media blaze.

Sokal consolidated his hoax with *Intellectual Impostures* (1997) in which he took on the entire French left-wing postmodern establishment.

Aunt Sallies akimbo!

Hoo-Whee! They're sitting ducks, pards!

Get 'em!

It was open season on Jacques Lacan, Julia Kristeva, Bruno Latour, Gilles Deleuze and Jean Baudrillard.

KNOCK 'EM DOWN - WIN A PRIZE!

Beyond the Hoax

Sokal's hoax proves what many radical and post-colonial critics of science already suspected.

The Public Understanding of Science

But we should not allow Science Wars, or the deep subjectivity of certain constructionists' positions, to distract us from the real issue: the power and authority, as well as the value-laden nature, of science.

The fury of the scientific establishment is based on an increasing realization that the traditional authority of science is rapidly eroding.

The legitimacy of science as the sole route to objectivity and truth has been damaged beyond repair.

THE ESTABLISHMENT

The hegemonic nature and values of science have been totally exposed.

Hence, the deep concern in scientific circles about the "public understanding of science".

Publicity vs. Accountability

The rubric "PUS" has been used to describe a continuum of activity. On one end, you have people, including some scientists, who see PUS as a public relations exercise and even a way of persuading audiences that controversial areas of science are unproblematic. On the other end of that continuum you have people, including scientists interested in public accountability, who want real dialogue about the future of research.

> After me... "Science is good for you".

> The public relations exercise tends to get more publicity, for obvious reasons, than the efforts at appealing to social responsibility and dialogue.

Under various PUS schemes, scientists are encouraged to learn communication skills so they can talk intelligently to the public. Journalists are encouraged to report science more accurately and widely.

> A brief restatement of our pledge not to inject any suppuration-centred humour around this nomenclature. Now... Back to bed.

How Science Has Changed

What both the scientists and the public have to realize is that science is changing drastically. It is changing not simply in the way the public perceives science – although this is one of the biggest changes experienced by science in the closing decades of the 20th century. The public has now realized that prejudice, fraud, professional jealousy and pride, and a desire for fame and fortune, are just as common amongst scientists as they are in the population at large.

> The changes in science are much deeper than simply those of perception ...

Practically, science is changing in the way it is <u>funded</u>, the way it is becoming <u>commercially driven</u>, and the way its <u>internal structures</u> are being transformed.

Conceptually, SCIENCE IS CHANGING BECAUSE UNCERTAINTY AND IGNORANCE ARE NOW CRUCIAL ELEMENTS OF ALL SCIENTIFIC ENDEAVOUR.

The Crux of Funding

The source of funding is perhaps the most obvious way values enter science.

Funding often influences the choice of problem to be investigated. If the funding is coming from a government source, then it will reflect the priorities of the government ...

> Whether space exploration is more important than the health problems of inner-city poor ...

> Whether nuclear power should be developed further or solar energy.

> The private-sector funding, mainly from multinationals, is naturally geared towards research that would eventually bring dividends in terms of hard cash.

After the Second World War, science funding in the US was dominated by three main players: government, industry and universities. The Federal government provided 50 to 60 per cent of total Research and Development (R&D) funding between 1953 and 1978. Over half of this funding was related to defence research. The funds went to universities and federally sponsored research institutions, and to public and private laboratories that further broadened Federal objectives, such as military security.

Corporate Funding of Research

After 1978, commercial funding for R&D began to exceed that of the government. By the early 1990s, corporations funded more than half of all research in the US. Industry expenditure on R&D is now two to three times the amount of Federal spending. Thus, most of the research done at the universities is now funded by industry.

Market and private sector imperatives now drive scientific and technological advances and determine what does and does not get funded.

This not only has serious implications for research ethics, accountability and conflicts of interest but makes science quite subservient to business interests.

Might I give you a cash injection, my dear?

The Profit Motive

Science is profit. And profit often determines the direction of science. The old military-industrial complex is being replaced by the corporation–university–private laboratory complex. Science becomes just another commodity, produced for sale.

What Direction for Science?

The marriage of science and profit can be detected in the major shift from physics to biology in the post-Cold War era. No private firm has ever supported a major particle accelerator, whereas the mapping of the human genome was eagerly propelled by private interests in both the US and UK.

There are no immediate profits to be made from discovering a new elementary particle.

But the human genome is an inexhaustible mine of innovative and marketable products.

... More caviar?

What Gets Scientific Attention?

Commercially driven science has two main characteristics. It focuses on certain areas of research at the expense of others; and it makes proprietorial claims on what most societies have regarded as "common knowledge" and what most individuals think is their intrinsic private property.

In general, this means that the problems of the Third World, where profits are limited, seldom get the attention of the researchers.

But since profits are associated with glamour, it also means that glamorous causes, usually those with celebrity endorsements, get serious attention.

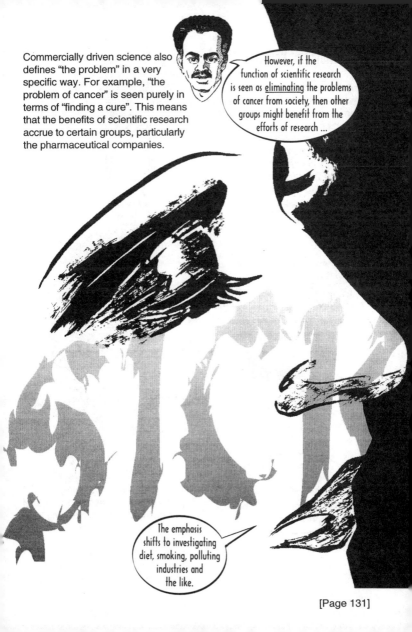

Population and Poverty

Similarly, the "problems of the developing countries" are measured in terms of "population". Research is focused on the reproductive systems of Third World women, methods of sterilization and new methods of contraception – all leading to Western products that can be sold to developing countries.

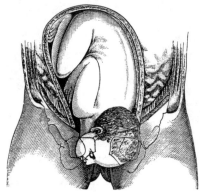

> However, if <u>poverty</u> were identified as the main cause of the population explosion, then research would take a totally different direction.

> Come on out, Hurry up! We need you to weed the millet patch.

The emphasis would have to shift to investigating ways and means of eliminating poverty, developing low-cost housing, basic and cheap health delivery systems and encouraging employment-generating (rather than profit-producing) technologies.

Patenting Knowledge

The commodification of science has produced a gold-rush system for patents. Anything that might conceivably have a use is now being patented, including the very stuff of life – sequences of DNA – as well as applied lab techniques.

One prominent scientist who produced a new "definition of life" was actually intending to patent his definition ...

But then I decided that that was God's work.

Oi... Watch it!

These new social problems of the abuse of science make the epistemological debates of "Science Wars" seem totally antiquated.

It is in the developing countries that the new predatory nature of science is most evident. Patenting of non-Western genetic resources began with the neem tree, as we'll see next.

The Neem Tree

Technically known as *Azadirachta indica*, the neem tree is a hardy, fast-growing evergreen tree that graces every village in the more arid regions of the Indian subcontinent. The *Upavanavinod*, an ancient Sanskrit treatise dealing with forestry and agriculture, describes how neem should be used for protecting plants from pests, curing ailing livestock and poultry and strengthening the soil.

> Various texts of Islamic yunnani medicine recommend neem as 100-per-cent effective contraception when applied intra-vaginally before intercourse.

Formulae are also given for making a whole range of medicines for such diseases as leprosy, ulcers, diabetes, skin disorders and constipation. Other texts have identified neem as a potent insecticide effective against locusts, brown plant-hoppers, nematodes, mosquito larvae, beetles and boll weevils.

Appropriation of Indigenous Knowledge

Commercially driven science is involved in patenting non-Western genetic resources, indigenous knowledge and ancient learning. Mexican beans, Filipino Jasmine rice, Bolivian quinoa, Amazonian ayahuasca, West Africa's sweet potatoes – all have been subjects of predatory intellectual "property claims".

> Everything from livestock germplasm to well-established medicines, indigenous knowledge of flora and fauna, even blood, is up for grabs.

> And the pirates are not just multinational corporations and government research organizations ...

Even respectable universities, along with individual scientific profiteers, are moving into indigenous communities under the guise of "research" – they then pilfer, patent and sell their "inventions" to larger enterprises.

Scientists used the local knowledge of farmers in Gabon to identify a particular variety of West African super-sweet berries. The active ingredient in the berries was then branded as a protein called "brazzein", said to be 2,000 times sweeter than sugar and thus an ideal candidate for a natural low-calorie sweetener. Between 1994 and 1998, four patents on the brazzein protein were obtained.

Intensified Appropriation

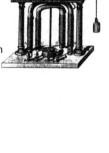

In some cases, entire indigenous systems are under assault. Over centuries, the Mayan communities in Mexico have developed a rich and sophisticated system of medical knowledge. Scientists use this system to guide their research. Interviews are conducted with Mayan "witch doctors" and "shamans", their herbal plants are collected and analysed, and their medical recipes scrutinized.

> Many of our plant products and medical processes have now been patented.

> Even blood cells from indigenous tribes have also been subject to patents.

The blood cell line of the Hagahia tribe in Papua New Guinea, which is infected with a virus that can lead to leukaemia, has been patented in developing a cure for leukaemia.

Mode 2 Knowledge

The total commodification of science, and its increasing domination by commercial and consumer interests, is also transforming science from within.

The conventional production of scientific knowledge, generated within the boundaries of a single discipline in cognitive context, is now being replaced by a new system. This new system has been called "*Mode 2 knowledge production*". In their seminal work, *The New Production of Knowledge* (1994), Michael Gibbons and his colleagues describe several attributes of knowledge production under Mode 2.

- Scientific work will no longer be limited to conventional institutions like universities, government research centres and corporate laboratories. There will be an increase in sites where knowledge will be created. Scientific work will also be done by independent research centres, industrial laboratories, think tanks and consultancies.

- These sites will be linked in various ways – electronically, organizationally, socially, informally – through functioning networks of communication.

- There will be simultaneous differentiation at these sites of fields and areas of study into finer and finer specialities. The recombination and reconfiguration of these subfields form the bases for new forms of useful knowledge.

"As a result, most scientists will become contract workers; they will work as temporary gangs of 'fungible' researchers, specially brought together to work on a particular problem and, at the conclusion of each project, redeployed or discarded. Researchers will become totally proletarianized as they lose their property, both in the skills of stable paradigm-based research, and in the rights to their results."

J. Ravetz

Consequences of Mode 2 Knowledge

"Mode 2" will be a radical departure from the types of social structures that science has had over the past centuries. Several emerging problems in these new social relations can be identified.
For instance …

- **What will ensure...**
  the preservation of the "academic" sector, still necessary for training and creativity, when it is inevitably assimilated into the new mode of knowledge production?

- **What will ensure...**
  the maintenance of quality-control, when the traditional informal "community" skills, etiquette and sanctions are rendered meaningless in a totally "commodity" enterprise?

- **What will ensure...**
  the survival of independence and criticism, when the management of troublesome elements does not need the crude threat of dismissal but only the subtler control of the blacklist?

- **What will ensure...**
  the recruitment of gifted young people, when the career image of independent searchers for knowledge is replaced by that of contract "geeks" in Mode 2?

Juh–... uh... just a minute, there... Wh– Who are you calling a geek, sir?

Uncertainty in Mode 2

Scientists have long known about uncertainty. Every time they start to investigate a problem, the possible answer is uncertain to some degree. But in normal science, the uncertainties are small; the puzzle is almost sure to be solved, and the possible answers are in a narrow range.

And although all the results in science have some uncertainty, they are mainly what we call "technical".

Figure 1 Successive accepted values of the fine structure constant a^{-1} (from BN Taylor et al 1969 *The Fundamental Constants and Quantum Electrodynamics* London: Academic p7)

Statistical methods can tame them, and they can be adequately expressed with "error bars".

Uncertainty occupies centre stage when **policy** is involved, and when **consumer-driven** science moves towards Mode 2 production of knowledge. Why does uncertainty become central?

Policy Debates in the Balance

In policy debates, uncertainties must always be balanced against "error costs". In the case of global warming, for example, some would suggest that the American economy must not be damaged by energy restrictions, unless we are quite sure about global warming.

Others would argue that, in spite of the remaining uncertainties, the dangers to humanity are clear.

In relation to uncertainty, science in the policy arena is therefore more like science in the law courts than like normal research science.

The value-commitments that actually shape all research are here quite open, explicit and contested. How uncertainty can affect policy was illustrated by the frightening case of "mad cow" disease.

"Mad Cow" Disease

"Mad cow" disease – or Bovine Spongiform Encephalopathy (BSE) – struck the UK in the 1980s as a strange epidemic of unknown causes, yet almost certainly related to intensive rearing and unnatural feeding practices (herbivorous cattle were fed on animal protein). As the epidemic spread, scientific advisers had to juggle the uncertainties of its ultimate economic cost, the price of control by mass slaughtering and the unlikely but still conceivable possibility of the disease spreading to humans.

In practice, the overriding concern seemed to be the welfare of the Ministry of Agriculture.

Even after cats caught the disease in 1990, there was official denial of danger to humans.

Measures of containment were all too little, too late and too partial. By 1996, when a human form of the disease was confirmed, there was brief general panic. The nation settled down to wait and see whether there would be isolated tragedies or mass horror.

The MMR Scare

We can see uncertainty in situations involving decisions about the control of ordinary infectious diseases. The UK Department of Health has a rigorous policy of simultaneous vaccination for three common childhood diseases: "MMR", or measles, mumps and rubella (chicken pox). Each of these can have severe effects on a minority of victims.

> But there is strong anecdotal evidence that the MMR vaccinations themselves are harmful – with risks of autism – if only to a very small minority of children.

> Official denials by the government have only aggravated the fears of many parents.

Epidemiological studies are rejected by critics as flawed. There is no consensus at all on the facts, and the values – the common good versus a risk of severe injury to *my* child – are in dispute. A large refusal of the "triple shots" would lead to a real danger of an epidemic of measles among the unvaccinated.

Assessing the Bigger Picture

In all such cases, the uncertainties go far beyond the merely "scientific". When planners are considering the threats of future floods (a likely consequence of global climate change), their decisions face the prospect of conflicts.

Preventing flooding upstream can increase the threat downstream.

There are threats to property values and businesses ...

... problems with insurance and assessing liability for past and future damage.

In all of these, uncertainties are severe, and the various interests can all too easily be set against each other.

Statistical Errors

The same level of uncertainty can be found deep within science. In any experiment involving statistical techniques, a choice is made between the errors of Type I (rejecting a true hypothesis) and of Type II (accepting a false hypothesis). Normally, the Type I errors are deemed to be more serious, and researchers automatically tune their tests accordingly.

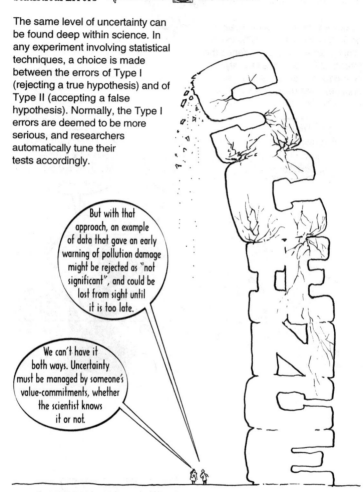

But with that approach, an example of data that gave an early warning of pollution damage might be rejected as "not significant", and could be lost from sight until it is too late.

We can't have it both ways. Uncertainty must be managed by someone's value-commitments, whether the scientist knows it or not.

Problems of managing uncertainty lead us to the question of ignorance.

The Place of Ignorance

Sir Peter Medawar (1915–87), British immunologist and Nobel Prize laureate

Science is the art of the soluble ...

It's a headache preparation?

This elegant formulation reveals much about the limits of scientific inquiry and its picture of the world.

For what is not soluble is not scientific. It does not count, it does not exist.

This hole does not exist!

Yoo-hoo!

This restricted view of science enhanced its power in the past. Now it presents perils for the future. To begin with, we are discovering that science seldom solves problems in neat packages – there are always extra bits that are not and cannot be solved. As in the case of the radioactive waste produced by nuclear power, these messy unsolved parts of the problem are typically neglected until they suddenly present crises in all dimensions.

The restriction of science to the "soluble" also has other, even deeper, effects on our vision of knowledge and the world. For it entails a total exclusion of *ignorance* from our view. Ignorance is not soluble by means of ordinary research. We therefore have no notion of its existence.

A Choice of Ignorance

Recognition of ignorance becomes very important for one very practical problem in scientific activity: *priorities* and *choices*. For whenever a proposed research project is given a low priority, it is not undertaken. As a result, the chance of gaining new knowledge is lost; and in that respect we remain *in ignorance*.

If our society is relatively less interested in – say – occupational health and alternative energy supplies than in hi-tech medicine and nuclear power, we remain in ignorance about those alternatives.

What we "know" is selected by these priorities and choices.

In this way, we can speak of ignorance as <u>politically</u> and <u>socially</u> constructed.

"By focusing on ignorance rather than on knowledge, we can escape some of the relativistic, sceptical implications of constructionist theories of science. It is easier for us to imagine ignorance as being conditioned by values and power."
J. Ravetz

"Ignorance-squared"

Ignorance of ignorance – or "ignorance-squared" – is a very recent phenomenon in European intellectual history. Continuously, from the time of Plato to that of Descartes, the ignorance of ignorance was a recognized category among philosophers. Socrates' quest was for awareness of his own ignorance. Ignorance was also an important concept in Islamic, Indian and Chinese science and philosophy. Renaissance humanist writers gave prominence to ignorance-squared as the worst intellectual failing.

> The big break came with Galileo and Descartes who imagined human knowledge as limitless in its scope and perfectibility.

> For us, ignorance is a void to be filled as quickly as possible.

> We each have a method whereby it is hoped and claimed that this can be accomplished.

The End of Doubt

Once Doubt had been conquered by Descartes, it hardly ever reappeared in the philosophy of science. But in our times, it has returned with a vengeance. In connection with speculative theories of cosmology, it is fun.

But ignorance is deadly serious when encountered in the selection of research and in gauging the dangers of proposed scientific innovations.

Modern science, with its myths of "objectivity", lacks the conceptual equipment to deal with ignorance-squared.

Uncertainty and ignorance of ignorance become pressing practical problems when safety becomes a major issue for science.

Safety and the Unknowable

Every advance in science ushers us towards new and hidden dangers. Consider, for example, how scientists assured the public that genetically modified crops were actually safer than those created by traditional processes. This was because scientists could directly alter the genes responsible for desired properties, leaving everything else untouched. Many of them really believed this, but it turned out to be false.

First, the insertion and activation of a new gene requires a severe disruption of the whole genetic machinery of the organism.

No one knows what collateral damage is done to the genome.

Then, since the "expression" of a gene results from a complex – and little understood – physiological process, the real effects of a new gene on the organism are unknowable.

Finally, what will happen in the environment as the new genes spread is a matter nearly of pure conjecture. We might get away with it for the first few crops, but then eco-catastrophe could strike at any time.

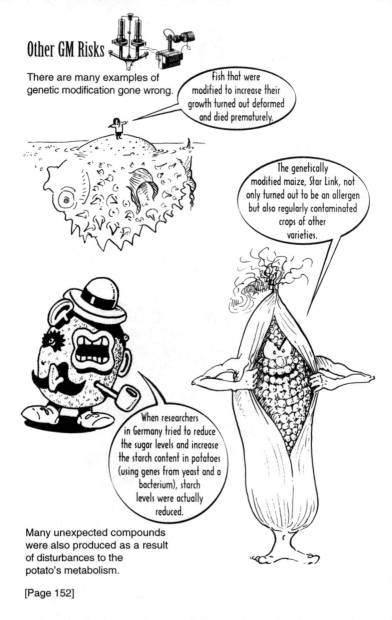

Increasing the Uncertainty Stakes

These isolated examples indicate the sorts of things that could happen, on an ever-increasing scale, as gene technology becomes established and routine. There is no way of knowing what sorts of harmful effects may occur; and some of them will certainly fail to be detected in standard safety checks. These cases, as well as the BSE ("mad cow" disease) crisis first in Britain then in Europe, show that our vast ignorance of possible harm is more important for policy than our limited knowledge of the possible pathways to that harm.

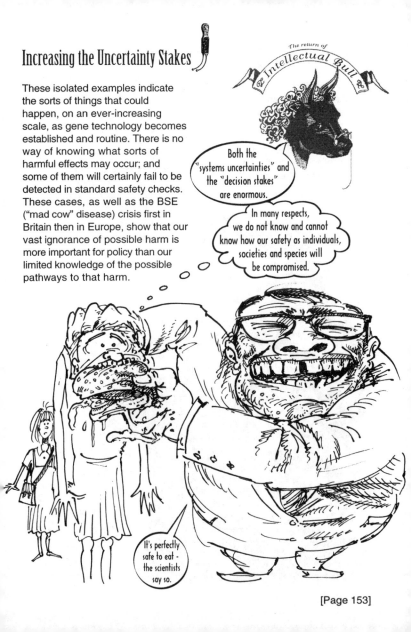

The return of Intellectual Bull

Both the "systems uncertainties" and the "decision stakes" are enormous.

In many respects, we do not know and cannot know how our safety as individuals, societies and species will be compromised.

It's perfectly safe to eat - the scientists say so.

Beyond the Normal

The combination of ignorance and uncertainty, as well as the practical changes to science – involving funding, commercialization, the complex issues of safety and new modes of knowledge production – all mean that science no longer functions in the "normal" way.

We find ourselves in a situation that is far from normal. Whenever there is a policy issue involving science, we discover that …

- Facts are uncertain.
- Values are in dispute.
- Stakes are high.
- Decisions are urgent.
- Complexity is the norm.
- Man-made risks may be running out of control.
- The safety of the planet and humanity is under serious threat.

We are thus moving into the era of post-normal science.

Post-Normal Science

Post-Normal Science (PNS) begins with the realization that we need a new style of science. The old image, where empirical data led to true conclusions and scientific reasoning led to correct policies, is no longer plausible.

The way forward must be <u>dialogue</u> based on the recognition of uncertainty and ignorance ...

Together with a plurality of legitimate perspectives and value-commitments.

Post-normal science is the sort of inquiry that occurs at the contested interface of science and policy. It can include anything from scientists' policy-related research to citizens' dialogue on the quality of that research.

Setting the Post-Normal Agenda

More specifically, post-normal science consists of a cycle of phases, constantly interacting, iterating and involving an agenda of issues.

* **Policy** – set in terms of general societal purposes, out of debate among the affected interests.

* **Persons** – who participates at any point, who selects them, by what criteria – and who selects the selectors?

* **Problem** – the defined task for the inquiry: recall that setting one problem excludes others and creates ignorance of the knowledge that they might have produced.

* **Procedures** – not just techniques, but also burden of proof: to what extent should absence of evidence of harm be taken as evidence of absence of harm?

* **Product** – who controls its management and diffusion, and who controls the controllers?

* **Post-Normal Assessment** – to what extent does the simple, tidy world of the laboratory or survey correspond to the complex, untidy world of policy and real experience?

In the arena of post-normal science …

- Scientific certainty is replaced by an extended dialogue.
- The "expert" is replaced by an "extended peer community" involving scientists, scholars, industrialists, journalists, campaigners, policy-makers and ordinary non-specialist citizens.
- "Hard facts" are replaced by "extended facts" which include not just published results but also personal experiences, local surveys and scientific information that was not intended for the public domain.
- Truth is replaced by Quality as the organizing principle.
- Scientific fundamentalism is replaced by the legitimacy of different perspectives and value-commitments from all those stakeholders around the table on a policy issue.

> "By changing the basic criterion from Truth to Quality, PNS relates the criticisms to practice, in all its dimensions from the technical to the ethical. It comprehends that Quality is not a simple attribute; indeed, it is **functional** (related to the use to which information will be put), **recursive** (who guards the guardians?) and **moral** (without an ultimate source of commitment, all quality collapses)."
>
> J. Ravetz

Post-normal science equips us all – scientists, citizens and decision-makers – with the tools necessary to deal with the complexities, uncertainties and risks inherent in contemporary science. It emphasizes the need to focus on the management of uncertainty and quality in making some of the most crucial decisions of our times. Conflict is not removed, but reconciliation based on understanding becomes possible.

In PNS, the ideal is not the attainment of some perfection in knowledge or practice but the <u>improvement of awareness</u> of oneself and of one's partners in a dialogue.

I feel that we're being watched.

Dialogue... Dialogue... Dialogue... Dialogue...

Ah, that's better.

PNS in Action

Post-normal science is now being realized in practice in many different ways. There is a growing number of citizens' science panels and consensus-orientated science conferences in Europe. Science centres are emerging and demand increasing for open public debates on various issues of science and society.

Patients' groups have an increasing role in determining strategies for managing their illnesses...

Sometimes – as in the case of AIDS – even negotiating on research methodology.

Such developments show that there are viable mechanisms for institutionalizing public participation in science.

With its core ideas of "extended peer community" and "extended facts", post-normal science includes the theories and campaigns connected with feminism, indigenous science and environmental justice.

More specifically, the principles of PNS can be seen in action in the "precautionary principle", community research networks and science shops.

The Precautionary Principle

The "precautionary principle", which recognizes the importance of uncertainty in the process and practice of science, is an indication of global recognition that science has become post-normal.

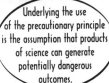

Underlying the use of the precautionary principle is the assumption that products of science can generate potentially dangerous outcomes.

We therefore need to proceed with caution.

BZZZT!
WAH-WAH!
CLANG! HONK!
Data overload... Data overload...
Emergency Code Red!!

The principle is now enshrined in many international regulatory statutes. When, and under what conditions, did the principle originate?

Origins of the Precautionary Principle

The classic formulation of the precautionary principle was first stated at the 1992 Climate Change Convention. There it was defined as "measures to anticipate, prevent or minimize adverse effects" of scientific progress "where there are threats of serious or irreversible damage". "Lack of full scientific certainty", the definition states, "should not be used as a reason for postponing such measures".

The definition even suggests that precautionary measures should be "cost-effective so as to ensure global benefits at the lowest possible cost".

The European Union's science policy is now guided by the spirit of the precautionary principle.

It is being used increasingly in policy-making in which there is risk to the environment or to the health of humans, animals or plants. The onus is now on the manufacturer to prove that a product or process is safe.

What does it matter? We'll die rich!

[Page 162]

Community Research Networks

Post-normal science insists that citizens must get involved in science. In the US, a number of vigorous Community Research Networks (CRNs) support non-profit and minority groups in their attempts to find solutions to problems of healthcare and pollution. Their activities are rooted in the communities they serve, and they encourage citizen participation at all levels. Examples of their work include …

Research to maintain jobs and environmental standards in the metalworking industry in Chicago, Illinois.

Helping communities assess the fairness of public-services distribution in Jacksonville, Florida.

Assistance in determining computer resources and access in Ohio neighbourhoods.

CRNs don't just bring science to the citizens; they encourage citizens to think scientifically about their problems.

The Community Responds ...

The affected families of Woburn responded by initiating their own epidemiological research.

> Eventually we were able to establish the existence of a cluster of leukaemia cases ...

> ... and then relate that evidence to industrial carcinogens leaked into the water supply.

> Our civil suit against the corporations responsible for the contamination resulted in an $8 million out-of-court settlement.

Woburn's victory gave major impetus for enacting federal Superfund legislation that provides resources for cleaning up the worst toxic waste sites in the US.

Science Shops

Science shops aim to provide independent participatory research support in response to concerns experienced by civil society. Their main function is to increase public access to and public awareness of science and technology.

Science shops initially developed in the Netherlands. Over the last two decades, a network of Dutch universities has set up dozens of science shops that conduct, co-ordinate and summarize research on social and technological issues in response to specific questions posed by community groups, public-interest organizations, local governments and workers.

Our shops are managed and operated by permanent staff and a regular supply of students who screen questions and refer challenging problems to university faculty members and research students.

The students get credits towards their degree for working at the shops and many do their postgraduate work on the problems brought to the science shops.

Many science shops have developed expertise in specific areas. Clients are often directed to the science shop best suited to address their particular concerns.

The Dutch system has, among other things, helped environmentalists to analyse industrial pollutants, workers to evaluate the safety and employment consequences of new production processes, and social workers to improve their understanding of disaffected teenagers.

The Dutch system has inspired science shops in Denmark, Austria, Germany, Norway and the Czech Republic.

Where Now?

The problems of understanding science are no longer focused on abstract questions of logic and knowledge. Those belonged to an earlier age, when Science, as the symbol of a secular society, was in conflict with Theology, the centre of a church-dominated social order.

Those who still proclaim "certainty" are either the survivors of the old triumphalist propaganda or the servants of the new arrogant corporations.

In the 21st century, science is a deeply contested territory.

Its increasing domination by private profit and corporate power cannot be masked any longer. Every advance in science encounters issues of uncertainty, ignorance, safety and control. The struggle now is over the shape and direction of scientific research and the control and use of its products.

The Democratic Solution

Science is the final frontier of democracy. It still aspires to "universal knowledge" yet remains in the hands of a self-selected few whose work is shrouded in "peer review" processes that occur beyond public scrutiny. Such elitism may have worked when science was still a form of gentlemanly pursuit that made few demands on the larger social and natural world.

Whose Science is It?

Science has become just too important to be left to the scientists and those who manage their work and control its products. Citizen participation at almost every level of the scientific enterprise has become essential.

This is needed both for maintaining the quality of policy-related science and for preserving democracy in a technology-driven age.

We seem to be getting older...

...and older...

Most of the fear and hostility that scientists sense in the public today does not come from ignorance or barbarism, but rather a feeling of disempowerment.

There is plenty of evidence – from book sales to television viewing – that ordinary people are in fact very interested and well-disposed towards science.

It is Our Science

There are indeed real issues to be explored by scholars in Science Studies, and real battles to be fought in a new Science Wars. But they are focused on sustainability, survival and justice. Science has at last entered the polity; it is no longer viable as "normal" puzzle-solving conducted in abstraction from the issues of *who pays and why*.

In this sense, we are in a post-normal age for science.

With this new, enriched awareness, science can regain its meaning for humanity and its reason for attracting the best talents to its endeavour.

Further Reading

Overviews
Ina Spiegel-Rösing and Derek de Solla Price (eds.), *Science, Technology and Society: A Cross-Disciplinary Perspective* (London: Sage, 1977); Colin A. Ronan, *Science: Its History and Development Amongst the World's Cultures* (New York: Facts on File, 1982); Sheila Jasanoff et al. (eds.), *Handbook of Science and Technology Studies* (London: Sage, 1995); Steve Fuller, *Science* (Buckingham: Open University Press, 1997); Mario Biagioli (ed.), *The Science Studies Reader* (New York: Routledge, 1999).

Politics of Science
Ziauddin Sardar, *Science, Technology and Development in the Muslim World* (London: Croom Helm, 1977); David Dickson, *The New Politics of Science* (Chicago: University of Chicago Press, 1986); Tom Wilkie, *British Science and Politics Since 1945* (Oxford: Blackwell, 1991); Sandra Harding (ed.), *The Racial Economy of Science* (Bloomington: Indiana University Press, 1993); Margaret Jacob (ed.), *The Politics of Western Science* (New Jersey: Humanities Press, 1994); Richard Sclove, *Democracy and Technology* (New York: Guilford Press, 1995); D.M. Hart, *Forced Consensus: Science, Technology and Economic Policy in the United States, 1921–1953* (Trenton: Princeton University Press, 1997); Jane Gregory and Steve Miller, *Science in Public* (Cambridge, MA: Perseus, 1998); Sheldon Rampton and John Stauber, *Trust Us, We're Experts* (New York: Penguin Putnam, 2001).

Philosophy of Science
Karl Popper, *The Logic of Scientific Discovery* (London: Hutchinson, 1959) and *Conjectures and Refutations* (London: Routledge and Kegan Paul, 1963); Paul Feyerabend, *Against Method* (London: NLB, 1975), *Science in a Free Society* (London: Verso, 1978) and *Farewell to Reason* (London: Verso, 1987); Imre Lakatos and Alan Musgrove (eds.), *Criticism and the Growth of Knowledge* (Cambridge: Cambridge University Press, 1970); J.R. Ravetz, *Scientific Knowledge and Its Social Problems* (Oxford: Oxford University Press, 1971) and *The Merger of Knowledge With Power* (London: Mansell, 1990).

Kuhn
Thomas S. Kuhn, *The Structure of Scientific Revolution*, (Chicago: University of Chicago Press, 1962); Barry Barnes, *T.S. Kuhn and Social Science* (London: Macmillan, 1982); and Steve Fuller, *Thomas Kuhn: A Philosophical History for Our Times* (Chicago: University of Chicago Press, 2000).

History of Science
George Sarton, *Introduction to the History of Science* (New York: Williams and Wilkins, 1947); J.D. Bernal, *Science in History* (Cambridge, MA: MIT Press, 1979); Joseph Needham, *Science and Civilisation in China* (Cambridge: Cambridge University Press, 1954–) and Ho Peng Yoke, *Li, Qi and Shu: An Introduction to Science and Civilization in China* (Hong Kong: Hong Kong University Press, 1985); D.M. Bose et al. (eds.), *A Concise History of Science in India* (Delhi: Indian National Science Academy, 1971) and Debiprasad Chattopadhyaya (ed.), *Studies in the History of Science in India* (Delhi: Asha Jyoti, 1992); Roshdi Rashed (ed.), *Encyclopaedia of the History of Arabic Science* (London: Routledge, 1996) and Donald R. Hill, *Islamic Science and Engineering* (Edinburgh: Edinburgh University Press, 1993); and Helaine Selin (ed.), *Encyclopaedia of the History of Science, Technology and Medicine in Non-Western Cultures* (Dordrecht: Kluwer, 1997).

Sociology of Science
Barry Barnes (ed.), *Sociology of Science* (London: Penguin, 1972) and *Scientific Knowledge and Sociological Theory* (London: Routledge and Kegan Paul, 1974); Ian Mitroff, *The Subjective Side of Science* (Amsterdam: Elsevier, 1974); Karin Knorr-Cetina, *The Manufacture of Knowledge* (Oxford: Pergamon, 1981); Bruno Latour and Steve Woolgar, *Laboratory Life: The Construction of Scientific Facts* (Princeton, NJ: Princeton University Press, 1986); Steve Fuller, *Social Epistemology* (Bloomington: Indiana, 1988); Harry Collins and Trevor Pinch, *The Golem: What Everyone Should Know About Science* (Cambridge: Cambridge University Press, 1993); Michael Gibbons et al., *The New Production of Knowledge* (London: Sage, 1994); Barry Barnes et al., *Scientific*

Knowledge: A Sociological Inquiry (London: Atholone, 1996).

Science and Empire
Daniel R. Headrick, *Tools of Empire* (Oxford: Oxford University Press, 1981); Michael Adas, *Machines as the Measure of Men: Science, Technology and Ideologies of Western Dominance* (Ithaca: Cornell University Press, 1989); Deepak Kumar, *Science and Empire* (Delhi: Anamika Prakashan, 1991) and *Science and the Raj* (Delhi: Oxford University Press, 1995); Roy MacLeod and Deepak Kumar (eds.), *Technology and the Raj* (London: Sage, 1995).

Feminist Critique
Sandra Harding, *The Science Question in Feminism* (Buckingham: Open University Press, 1986); Maureen McNeil, (ed.) *Gender and Expertise* (London: Free Association Books, 1987); Hilary Rose, *Love, Power and Knowledge* (Oxford: Polity Press, 1994); Margaret Wertheim, *Pythagoras' Trousers* (London: Fourth Estate, 1997); Jean Barr and Lynda Birke, *Common Science?: Women, Science and Knowledge* (Bloomington: Indiana University Press, 1998).

Post-colonial Critique
Ziauddin Sardar (ed.), *The Touch of Midas: Science, Values and the Environment in Islam and the West* (Manchester: Manchester University Press, 1982), *Explorations in Islamic Science* (London: Mansell, 1985) and *The Revenge of Athena: Science, Exploitation and the Third World* (London: Mansell, 1988); Ashis Nandy (ed.), *Science and Violence* (Delhi: Oxford University Press, 1988) and Claude Alvares, *Science, Development and Violence* (Delhi: Oxford University Press, 1992); Sandra Harding, *Is Science Multi-cultural?* (Bloomington: Indiana University Press, 1998).

Science Wars
"Science Wars", *Social Text*, vol. 46–47 (Durham: Duke University Press, Spring/Summer 1996); Paul R. Gross et al. (eds.), *The Flight From Science and Reason* (New York: New York Academy of Science, 1996); Paul Gross and Norman Levitt, *Higher Superstition* (John Hopkins University Press, 1994); Alan Sokal and Jean Bricmont, *Intellectual Impostures* (London: Profile Books, 1997); Thomas Gieryn, *Cultural Boundaries of Science: Credibility on the Line* (Chicago: University of Chicago Press, 1999); Ziauddin Sardar, *Thomas Kuhn and the Science Wars* (Cambridge: Icon Books, 2000).

Post-Normal Science
Silvio Funtowicz and J.R. Ravetz, *Uncertainty and Quality in Science for Policy* (Dordrecht: Kluwer, 1990); J.R. Ravetz (ed.), "Post-Normal Science", Special Issue of *Futures*, vol. 31, September 1999; Hilda Bastian, *The Power of Sharing Knowledge: Consumer Participation in the Cochrane Collaboration*, http://www.cochraneconsumer.com.

About the Author and Artist

Ziauddin Sardar, a renowned cultural and science critic, is a pioneering writer on Islamic science and the future of Islam. A visiting professor of post-colonial studies at the City University, London, he has published over 30 books on various aspects of science, cultural studies, Islam and related subjects, many of which have been translated into over 20 languages. Professor Sardar is the editor of *Futures*, the journal of policy, planning and futures studies. His most recent books include *Postmodernism and the Other* (1998), *Orientalism* (2000) and *Aliens R Us: The Other in Science Fiction Cinema* (2001), which he has co-edited with Sean Cubitt, and *The A-Z of Postmodern Life* (2002). He has also written guides to Muhammad, cultural studies, chaos, media studies and, with Jerry Ravetz, mathematics, in the *Introducing* series.

Borin Van Loon has illustrated *Darwin and Evolution, Genetics, Buddha, Eastern Philosophy, Sociology, Cultural Studies, Mathematics, Media Studies* and *Critical Theory* in the *Introducing* series. He snips away at piles of cuttings in an obsessive sort of way and is confirmed in his opinion that "*Introducing* Books are the Mother of Invention".

Acknowledgements

We would like to thank Gail Boxwell for her invaluable support.

Index

academic left 114

Big Science 56–7
Bloor, David 63

Carnap, Rudolf 44
certainty *see* uncertainty
Chinese
  scientific knowledge 89, 91
  view of nature 99
Church, the 23–4
Civilizational Science 92
climate change 142, 145, 162
colonial science 87–9
commercially driven science 126–39
Community Research Networks (CRNs) 164–7
Conant, James Bryant 57
constructionists 65–73, 158

Darwinism 102
democracy 171
Descartes, René 34, 149
doubt 150

education and science 110
Einstein, Albert 56
environmental movement 39
epistemology *see* social epistemology
error costs 142, 146
experiments 10–11

falsifiability 46
feminist criticism 72
Feyerabend, Paul 58–60
Fuller, Steve 15, 106

Galilei, Galileo 33, 41–2
Galton, Francis 12
genetic
  difference 102
  modification 151–3
Gibbons, Michael 139
Gross, Paul 114–15

Haber, Fritz 38
Harding, Sandra 77–80
Hinduism 100

ignorance 147–150, 155
Indian science 91, 96–7
indigenous knowledge, using 133–8
induction 47
infectious diseases 143–4
Islam 89–90, 93–4, 111

Kuhn, Thomas 40–1, 48–57
  opposition to 53

Latour, Bruno 66–7
Lavoisier, Antoine 42
Leibniz, Gottfreid Wilhelm 42
Levitt, Norman 114–15
logical positivism 44–5

masculine thought 72
media and science 123
medical research 130–1, 134
Mendel's theory 71
Merton, Robert 62
Mitroff, Ian 67
Mode 2 knowledge production 139–41

nature 98–9, 100
Neurath, Otto 44
Newton, Isaac 42
non-violent communication 95
normal science (Kuhn) 50–2

objectivity 4–5
observations and tradition 71

paradigms, Kuhn 49–51
patents 133
pollution 165–7, 169
Popper, Karl 46–7, 53
positivism 44–5
post-colonial scientists 84–6, 90
postmodern science 158
post-normal science 154–60
poverty and commercial science 132
power imbalances 81
precautionary principle 161–3
Priestley, Joseph 42
profit and science 127
public participation 159–60, 164–73
Public Understanding of Science (PUS) 121–2

racial bias of science 103–5
reality 65–6, 69
reproductive labour 83–4
research
  and development 126
  and ignorance 147–9
  medical 130–1, 134
revolutionary science 51

safety 150–4
Schlick, Moritz 45

science
  attempt to define 4, 14
  changing 124
  defended 113
  misunderstood 18
  public disenchantment 112
  racial inequality of 103–5
  revolutionary 51
  shops 168
  see also Big Science
Science Studies
  aims 31
  background 20
  defined 19
  different approaches 21
  importance of 30
  school of 22
Science Wars 116–20
scientists, our view of 6–9
Scientology of Scientific Knowledge (SSK) 61, 64
sexual prejudice *see* women and science
Snow, C.P. 63
social epistemology 106–12
Social Text 116
Sokal, Alan 116–19
Strong Programme, The 63–4, 69

theories, evolving 71
trust in science, lack of 6

uncertainty 141–6, 153, 170

Vienna Circle 44–5

Western
  bias 103–5
  science 89
  view of nature 98
women and science
  invisible in lab 75
  segregation 74
Woolgar, Steve 66–7